Logistics for Engineers

In today's rapidly evolving technological landscape, the field of logistics has undergone a significant transformation from a conventional supply chain operation to a highly intricate, data-driven network that is crucial for the efficient functioning of contemporary society. This new textbook seeks to facilitate this transition by providing a framework rooted in engineering fundamentals, empowering students to navigate the complexities of the modern logistics industry.

Logistics for Engineers: Automatic Control, Network Science, and Industry 4.0 provides engineering students with a comprehensive and practical approach to mastering logistics by incorporating principles from Automatic Control, Network Science, and Industry 4.0. This new textbook discusses the intricate world of logistics as a complex interconnected system, where inventory levels, supply chains, and production processes act as signals within a control system. By integrating fundamental mathematical models such as differential equations and Laplace transforms, along with industry-specific applications, this textbook empowers future engineers with the necessary tools to optimize efficiency, reliability, and responsiveness within modern, data-driven logistics networks. Through a combination of real-world examples, challenging problem-solving exercises, and insights into cutting-edge technologies, this textbook prepares readers to excel in today's fast-paced industrial environment.

Designed specifically for undergraduate engineering students, this textbook is tailored to those studying Industrial Engineering, Mechanical Engineering, Systems Engineering, and Operations Engineering. It serves as a foundational resource for courses in logistics, supply chain management, and operations engineering. The content is particularly beneficial for graduate students focusing on logistics, supply chain analytics, and smart manufacturing, as it offers a thorough exploration of logistics processes using a mathematical and systems-based methodology.

The textbook provides a variety of resources to enhance learning, including end-of-chapter assessments, exercises, case studies, and figure slides.

Logistics for Engineers
Automatic Control, Network Science, and Industry 4.0

Javier Villalba-Diez
and Joaquín Ordieres-Meré

CRC Press
Taylor & Francis Group
Boca Raton London New York

CRC Press is an imprint of the
Taylor & Francis Group, an **informa** business

Designed cover image: Shutterstock

First edition published 2026
by CRC Press
2385 NW Executive Center Drive, Suite 320, Boca Raton FL 33431

and by CRC Press
4 Park Square, Milton Park, Abingdon, Oxon, OX14 4RN

CRC Press is an imprint of Taylor & Francis Group, LLC

ISBN: 978-1-041-04213-6 (hbk)
ISBN: 978-1-041-04343-0 (pbk)
ISBN: 978-1-003-62790-6 (ebk)

DOI: 10.1201/9781003627906

Typeset in Nimbus font
by KnowledgeWorks Global Ltd.

Publisher's note: This book has been prepared from camera-ready copy provided by the authors.

Access the Instructor and Student Resources: Routledge.com/9781041042136

"It is difficult to give understanding to someone who lacks will, and even more so to give will to someone who lacks understanding."

Balthasar Gracian, 1647

Contents

Foreword

It is a pleasure to introduce this textbook on the application of advanced mathematical and control-theoretic methods in modern logistics and supply chain management. In a world shaped by digitalization, globalization, and increasing volatility, logistics has evolved into a highly complex business and engineering discipline. This book provides students with a rigorous, mathematically grounded roadmap for analyzing and optimizing the dynamic systems that sustain today's supply chains.

At Hochschule Heilbronn, we are committed to bridging theory and practice. Every day, we witness how Industry 4.0, sustainability, digitalization, and artificial intelligence—ranging from autonomous warehouses and IoT-enabled tracking to data-driven forecasting and green routing—are transforming logistics. To effectively harness these innovations requires not only domain expertise but also a firm command of analytical tools such as differential equations, Laplace transforms, transfer functions, and network spectral theory.

This textbook successfully builds that foundation. Its structure—progressing from systems thinking, through mathematical modeling and control, to case studies and emerging technologies—reflects the natural journey of a logistics engineer and manager. Classical methods are integrated with control frameworks such as PID, model predictive control, and adaptive strategies. Traditional lean manufacturing concepts, such as Kanban, are insightfully reinterpreted as dynamic feedback systems.

Equally noteworthy is the treatment of supply chains as networks. By modeling them as graphs with adjacency and Laplacian matrices, students gain tools to evaluate connectivity, robustness, and sustainability. This systems perspective is essential for designing resilient, circular, and green supply chains that can withstand disruptions and shocks.

What distinguishes this work is its balance of rigor and relevance. Theoretical developments are consistently reinforced with examples, simulations, and exercises, making it an invaluable resource for both teaching and self-study. I am confident this textbook will become a cornerstone of logistics education. It not only reflects the state of the art in academic research but also addresses the urgent needs of industry. By equipping students with analytical skills and systems-level intuition, it prepares them to meet the multifaceted challenges of twenty-first-century logistics.

I warmly congratulate the authors and commend this work to educators, students, and professionals. May it inspire the next generation of logistics engineers and managers to innovate responsibly and contribute meaningfully to a sustainable global economy.

Prof. Dr. Ralf Dillerup
Dean, Business School
Hochschule Heilbronn
August 2025

Preface

The modern field of logistics engineering stands at the nexus of operations research, systems theory, and advanced mathematical modeling. Over the past decades, global supply chains have evolved into highly interconnected and dynamic networks, demanding analytical rigor and engineering precision. This book has been conceived to equip bachelor-level students in engineering, operations research, and related disciplines with a systematic, mathematically grounded understanding of logistics as a complex, networked control system.

Our motivation derives from the observation that traditional logistics textbooks often treat inventory management, transportation planning, and facility location as distinct problems, solved with ad hoc heuristics. In practice, however, these elements interact continuously: inventory levels influence replenishment rates; lead times shape order-up-to policies; and network topologies constrain flows and resilience. To capture these interactions, we adopt a unifying perspective that models inventory, flows, and network structure as mathematical signals and operators. Ordinary and partial differential equations, Laplace transforms, transfer functions, and spectral graph theory become the lingua franca by which we analyze stability, responsiveness, and robustness.

The book is organized in three main sections:

- **Section I: Introduction** We begin with Chapter 1, which frames logistics as an engineering discipline. Core concepts—including system boundaries, performance metrics, and the role of technology—are introduced alongside a systems approach to design and optimization. This chapter lays the groundwork for viewing logistics problems through the lens of continuous-time dynamics and control theory.
- **Section II: Mathematical Models in Logistics** Chapters 2 through 6 develop the analytical machinery. Chapter 2 presents classical ODE models, Laplace-domain methods, and their application to single- and multi-node inventory systems. Chapter 3 recasts inventory management as a feedback control problem, examining transfer functions, the bullwhip effect, and multi-echelon dynamics. Chapter 4 extends these ideas to automatic control—particularly PID, MPC, adaptive, and fuzzy controllers—in logistics contexts. Chapter 5 focuses on pull-based Kanban systems, quantifying responsiveness, reliability, and the interplay of push vs. pull policies. Finally, Chapter 6 introduces network science: graph topologies, spectral properties (Laplacian eigenvalues, Fiedler vectors), and resilience metrics, culminating in sustainable manufacturing and circular supply-chain models.
- **Section III: Case Studies, Applications, and Trends** Chapters 7 and 8 bring theory to life. Through case studies—from automotive and pharmaceutical supply chains to e-commerce and green routing—we illustrate how

mathematical models inform real-world decisions. The final chapter surveys emerging trends: IoT, robotics, blockchain, and smart factories—pointing toward future research directions.

Throughout the text, we emphasize:

1. **Signal-Based Thinking**: Inventory and demand are treated as time-varying signals; replenishment policies become control laws.
2. **Interconnected Systems**: From single-node ODEs to large-scale network Laplacians, we demonstrate how local policies propagate through the supply chain.
3. **Mathematical Rigor and Practical Insight**: Each analytical development is accompanied by numerical examples, simulations, and design guidelines, ensuring that students appreciate both the theory and its logistical implications.

This volume assumes familiarity with calculus, linear algebra, and basic probability. For computational exercises, students may use MATLAB, Python (SciPy, NetworkX), or similar platforms.

It is our hope that readers will emerge from this book not only with a robust toolkit of analytical methods but also with a systems-level intuition for designing resilient, efficient, and sustainable logistics networks. As supply chains continue to evolve under the influences of globalization, digitalization, and sustainability imperatives, the mathematical foundations presented here will serve as a solid platform for both academic research and industrial innovation.

We extend our gratitude to the many colleagues and graduate students whose insights shaped this manuscript and to the practitioners whose challenges inspired our models. May this work foster the next generation of logistics engineers who will tackle the complexities of twenty-first-century supply chains with analytical rigor and creative vision.

Dr. Javier Villalba-Diez
Dr. Joaquín Ordieres-Meré

About the Authors

Dr. Javier Villalba-Diez is a distinguished expert with the unique accomplishment of holding two PhDs from the Universidad Politécnica de Madrid (UPM)—both awarded for excellence as the best doctoral theses in their fields. His first PhD in Engineering, Economics, and Organizational Innovation (2016) focuses on Strategic Organizational Design, while his second PhD in Quantum Computational Applications for Industry 4.0 (2022) underscores his cutting-edge expertise in future-ready technologies.

Driven by a mission to empower individuals and organizations to achieve strategic goals while building trust, Dr. Villalba-Diez is a Mechanical Engineer from Technische Universität München and an Industrial Engineer from UPM (2003). Over the past 20 years, he has served as a professor, consultant, and production manager in German, American, and Japanese manufacturing environments, applying his deep technical and strategic knowledge to real-world challenges.

His research interests span transformative areas such as Quantum Computing, Deep Learning, Hoshin Kanri, Strategic Organizational Design, and Business AI. Dr. Villalba-Diez's work takes him around the globe, collaborating with companies and visiting hundreds of factories to test innovative ideas and share valuable insights.

With his dual PhDs and extensive global experience, he continues to push the boundaries of industry and technology, dividing his time between Germany, the USA, Japan, and Spain.

Dr. Joaquín Ordieres-Meré earned the Ph.D. degree in industrial engineering from the Universidad Nacional de Educación a Distancia (UNED) in 1987. He was a full professor of industrial management with the Universidad de la Rioja in 1997 and has been with the Universidad Politécnica de Madrid since October 2008, where he continues to contribute to the field of engineering and its applications in industry. His research expertise lies at the intersection of business analytics, with a particular focus on leveraging data-driven modeling to enhance knowledge discovery and optimize complex processes. This aligns seamlessly with the principles of logistics engineering, where mathematical rigor and analytical tools are crucial for designing efficient and resilient supply chains. Throughout his career, Dr. Ordieres-Meré has been deeply involved in over 70 research projects, many of which are international and highly competitive. His extensive contributions are reflected in more than 150 research papers, accumulating over 7700 citations, demonstrating a significant impact on the academic and practical understanding of industrial and logistics systems. He actively participates in ISO TC groups, contributing to the standardization and best practices within the industry. His commitment to academic discourse is evident in his regular service as a reviewer for various esteemed journals, including his roles on the Editorial Boards of MDPI Sensors, the International Journal of Data Mining, Modeling, and Management, and Frontiers in Built Environment. Furthermore, he represents his country as a member of several European Union expert committees, such as RFCS TGA5, further cementing his influence in shaping the future of industrial and logistical landscapes. His work consistently emphasizes the application of mathematical and analytical tools to tackle complex logistics challenges, from optimizing inventory and transportation to fostering robust and sustainable supply chain operations.

Section I

Introduction

1 Introduction to Logistics as an Engineering Discipline

1.1 INTRODUCTION

The field of logistics has grown significantly from a supporting function to a core element of modern engineering systems [25, 17]. This section will introduce logistics as an engineering discipline, define the term, trace its historical development, and outline the relevance of the subject in modern industrial systems [18, 20].

1.1.1 DEFINITION OF LOGISTICS

Logistics is broadly defined as the art and science of planning, implementing, and controlling the efficient flow of goods, services, and information from point of origin to point of consumption [7, 19]. This definition encapsulates key concepts such as efficiency, integration, and optimization that are foundational to logistics engineering.

In simple terms, logistics is the coordination of movement and storage, which includes goods, services, and information. It is a discipline that heavily relies on planning and implementation; it requires highly developed systems in order to manage resource flow. According to Ghiani et al. [10], this view emphasizes the intrinsic relationship between logistics, design, and systems engineering, laying strong emphasis on technological and analytical approaches.

Logistics is, however, not a static blueprint, and it thrives in the dynamism of real-world operations; logistics involves responding in an agile way to the changes, necessitating detailed coordination of people, facilities, and supplies against the challenge at hand. The concept underlines another crucial attribute that relates to the core elements of decision-making and flexibility of logistics systems operating within unpredictable and uncertain conditions of contemporary relevance.

Furthermore, logistics involve not only the flow of goods physically but also those from the realm of intangibility: services. It requires elaborate design and management and the perpetual process of improving the service networks that facilitate undisturbed product delivery and effective fulfillment of services.

Finally, the historical origins of logistics should be noted to lie in the military realm, where the function was considered to be a science for many years in the planning and implementation of moving and sustaining forces. In this view, the ability of logistics to deliver operational preparedness and efficiency in austere and sometimes unexpected environments is foregrounded.

According to the explanation of Bowersox [4], in engineering, logistics transcends the traditional functions of transport and storage by the application of mathematical

DOI: 10.1201/9781003627906-1

modeling, system design, and technology in order to increase efficiency. This therefore means that logistics engineering includes the elimination of inefficiency along the supply chain with an emphasis on reducing environmental impact and improving sustainability [11]. It applies systems thinking in the optimization of material and information flows as an advanced tool.

1.1.2 OVERVIEW OF LOGISTICS ENGINEERING

Logistics engineering is one of the most representative and specific interdisciplinary scientific sectors to which engineering principles are applied to the design, optimization, and management of logistics systems. It embodies not only traditional logistics but also a system approach aimed at enhancing the effectiveness, dependability, and sustainability of supply chain operations. In essence, logistics engineering encompasses elements of industrial engineering, systems engineering, operations research, and information technology in an integrative framework at the physical, informational, and financial flows within and across organizations.

First, logistics engineering has a problem-solving focus and also views the issues as related systems. The logistics system is a very complex interrelation between networks of transportation, warehousing, inventory, and customer service, among other things. These varied components should integrate well to make the broader goals of cost efficiency, responsiveness, and sustainability possible. For example, transportation route optimization, one of the most typical problems in logistics engineering, involves highly complex algorithms that are variations of the Vehicle Routing Problem (VRP) or the Traveling Salesman Problem (TSP), which have been under study in the scope of operations research [9].

An indicative example of how logistics engineering can potentially change the world is the introduction of automated guided vehicles in warehouses. Companies like Amazon have utilized robotics and automated systems to optimize order picking and inventory replenishment processes. Such implementations follow the principles of logistics engineering through mathematical modeling and simulation, which provide a forecast for system performance and locations of bottlenecks [28]. Apart from the enhanced operational efficiency, these technologies show how logistics engineering can be applied to real-time decision-making and flexibility.

Despite such developments, a number of major challenges are encountered by logistics engineering. First is the growing international supply chain complexity. Drivers such as globalization and e-commerce require logistics networks to handle a diverse supplier base, longer lead times, and dynamic consumer demand patterns as depicted in Figure 1.1. This creates conflicting trade-offs among cost, speed, and service quality, which are not easy to balance. For example, JIT systems are indeed efficient under stable conditions but are quite vulnerable to disruption, as was evidenced by the COVID-19 pandemic, which severely interrupted supply chains.

Another very important challenge is sustainability. In particular, transportation of goods widely contributes to gaseous emissions. Logistics engineering should take note and come up with ways that will reduce environmental degradation by developing green logistics practices and adopting technologies such as electric vehicles and

Figure 1.1 Mosaic of Logisitic situations fueled up by Engineering and Technology.

renewable energy. Researchers in green supply chains claim that closed-loop systems and circular economy principles help reduce impacts [11].

Logistics engineering also plays a vital role in the rise of urban logistics and last-mile delivery, which have become critical issues with the growth of e-commerce. Companies are looking to innovative solutions such as micro-fulfillment centers, drone delivery systems, and crowd-sourced delivery models to solve challenges related to urban congestion and customer expectations. For example, Walmart's investment in micro-fulfillment technologies shows how logistics engineering can adapt to evolving consumer demands and urban infrastructure constraints.

Moreover, technological development gave a prodigious contribution to the industry. AI, IoT, and blockchain have become the most powerful tools for transformation in logistics engineering today. With AI-driven predictive analytics, companies can predict demand and keep their inventory at optimal levels, while IoT devices give real-time visibility into supply chain operations. Contrarily, blockchain allows for a secure and transparent record of supply chains to trace transactions and goods. The case of Maersk partnering with IBM in the development of a blockchain-based TradeLens platform perfectly shows how logistics engineering integrates such technologies into developing better transparency and efficiency.

The adoption of such advanced technologies also faces their own challenges in terms of implementation costs, data privacy concerns, and skilled personnel. These challenges have to be dealt with by logistics engineers while they ensure that the technological solution would fall in line with the goals of the organization and yield measurable benefits. Moreover, greater use of digital systems exposes the logistics operations to the risk of cyberattacks, requiring sound cybersecurity measures that can protect the critical infrastructure from damage.

In general, logistics engineering is basically a dynamic and important field and acts as an essential core for modern supply chains. Solving complex logistics problems with innovation in various industries applies scientific rigor with practical application. New challenges, at the same time, are created in each new set of emerging issues: sustainability, urbanization, and technological disruption. Logistics engineering will continue to play a vital role in global commerce and industrial development as innovative solutions using advanced analytics are developed and applied.

1.2 FUNDAMENTALS OF LOGISTICS ENGINEERING

The very foundation of logistics engineering should be laid down in its purpose: to design, optimize, and manage logistics systems by the application of principles of engineering, analytic tools, and technology. Logistics engineering bridges the chasm between theory and practice and facilitates smooth flow across supply chains: goods, information, and finance. It contrasts with traditional logistics because its very core is in the system of problem-solving, optimization, and integration of processes, paying close attention to precision, scalability, and adaptability to the management of complex systems.

1.2.1 CORE CONCEPTS AND PRINCIPLES

Some of the really important concepts are at the heart of the engineering that is done in logistics:

- systems thinking,
- flow management, and
- optimization.

Systems thinking is perhaps the first because logistics operations are part of a larger system with connected parts. One part—transportation, for example, or warehousing, or inventory management—can have an impact on another part; often, a system can be optimized in ways surpassing optimization of the individual components. For instance, improving warehouse storage efficiency might be counterproductive if it disrupts the flow of goods to the distribution centers and causes delays downstream. According to Bowersox et al. [5], this could result in lost business.

The second cornerstone is flow management, which concerns the smooth flow of goods and information. This involves minimizing waste, such as bottlenecks, idle time, and overproduction, while ensuring responsiveness to customer needs. It does this through the use of process mapping and simulation modeling tools that allow engineers to conceptualize the flow of resources and even simulate scenarios before actual execution. The just-in-time methodology best describes this principle by synchronizing production schedules with demand, hence reducing inventory levels and increasing efficiency, though at the expense of managing variability [14].

Optimization arguably is the most defining principle in logistics engineering. It involves the application of mathematical models and algorithms in pursuit of the optimum solutions to logistical problems. Classic problems, such as the Traveling Salesman Problem or the Vehicle Routing Problem, are extensively studied in operations research and provide basic models for logistics optimization. These models enable decision-makers to design cost-effective routes, minimize fuel consumption, and improve delivery times. For instance, logistics companies like FedEx use optimization algorithms that allow them to manage huge delivery networks and operate efficiently and on time. It was said by Ghiani et al. [9].

Data-intensive and analytics-based: This is another critical characteristic of logistics engineering. The modern logistics system operates on a vast amount of data coming from sensors, tracking devices, and transactional systems. In fact, such data in real-time enable predictive and prescriptive decision-making in building agile and productive logistics networks. For example, predictive analytics, through the power of artificial intelligence, can help a firm to forecast demand and position the inventory accordingly in order to avoid disruption. This idea is further supported by the integration of the Internet of Things, which facilitates real-time monitoring and coordination. It provides the capability for supply visibility throughout the chain, as Waller and Fawcett [28] confirm.

Besides these basic principles, logistics engineering also has its specific design methodologies. Systems engineering is one of the most widely used approaches, emphasizing definition of requirements, system design, implementation, and validation. This will give a systematic procedure to develop solutions that fit into the objectives and constraints of organizations. The design of a distribution network would normally require a step in identifying the customer demand pattern, finding optimal locations for facilities, and modeling the transportation route. Such designs are typically validated through simulation and pilot testing to mitigate risks before full-scale deployment.

It also put to the test the basics of logistics engineering while being robust in methodology. Often enough, uncertainties and variability in demand, times of lead, and conditions of transport complicate this optimization process. In essence, this is an indication that the additional components to logistics engineering concern risk management and contingency planning, as indicated in Figure 1.2. Such is the disruption that came from the COVID-19 pandemic, proving how fragile the global supply chains could get, asking for more resilience within logistics systems.

Moreover, sustainability has emerged as a critical concern in logistics engineering. Traditional optimization metrics, such as cost and time, are increasingly supplemented by environmental considerations. Green logistics focuses on minimizing the carbon footprint of logistics operations through practices such as route optimization, energy-efficient technologies, and the adoption of renewable energy sources. Circular supply chains, which integrate recycling and reuse into the logistics process, represent a paradigm shift toward sustainability [11].

Practical applications of logistics engineering further illustrate how the principle works. For instance, urban logistics systems' design would present challenges to be resolved in the context of traffic congestion, last-mile delivery, and infrastructural constraints. Companies like UPS and DHL have already developed

Figure 1.2 Engineering approach to Risk management in Logistics.

micro-distribution centers and used alternative methods of delivery, such as drones and electric bikes, in order to perform urban delivery efficiently. These examples show the adaptability of logistics engineering to various contexts without losing its gaze on efficiency and innovation.

Fundamentally, the basis of logistics engineering lies in its systematic approach to managing complex systems in logistics. Systems thinking, flow management, and optimization combined with advanced analytics and sustainability practices bring forth solutions for various complex issues of today's supply chains. These are not just principles but are being continually applied, tested, and updated to spur innovation and resiliency across a connected world.

1.2.2 KEY COMPONENTS OF LOGISTICS SYSTEMS

The components of logistic systems represent the structural and functional elements that enable the effective and efficient flow of goods, information, and resources across supply chains. Logistics engineering provides the theoretical and practical tools necessary to design, optimize, and integrate these components into a cohesive system based on several central components, such as:

1. Transportation Management,
2. Inventory Management,
3. Warehousing and Storage Systems,
4. Distribution Network design,
5. Information and Communication Technology,
6. Reverse Logistics, and
7. Human Resources and Organizational Systems.

Transportation management is central to any logistics system, which enables the physical movement of the goods. Transport is not only a conveyance mechanism; it is also a strategic aspect that influences cost, service level, and operational flexibility. Logistics engineers approach transportation by optimizing networks through routing algorithms, vehicle utilization studies, and modal selection frameworks. For instance, FedEx uses sophisticated routing software that dynamically assigns deliveries to vehicles to reduce fuel consumption and enhance the reliability of its services [5]. Such initiatives are part of a wider logistics engineering that aims at synchronization between transportation and other elements.

Inventory management is closely related to transportation and is the guide on the flow of goods within supply chains. Inventory is used both as a buffer against uncertainties and as a major source of costs. The dual nature of this aspect is balanced by logistics engineering using mathematical models for Economic Order Quantity and probabilistic demand to set an optimum level of stock so that the holding cost will be minimized. Inventory management practices popularized by Toyota are based on a 'just-in-time' philosophy whereby inventories are aligned with the production schedule to minimize any type of waste [27].

The other critical components of the whole network are the warehousing and storage systems, acting as nodes in the system of consolidation, sorting, and distribution. Warehousing operations consist of complicated decisions on layout design, material handling, and automation, which are all considered as part of logistics engineering. Many large, contemporary warehouses have incorporated robotics, conveyor systems, and real-time tracking in a race for improved productivity. This is evident from Amazon's state-of-the-art fulfillment centers using logistics engineering principles. For instance, AGVs and advanced tracking of inventories not only accelerate the order processing but also optimize the space utilization [28].

The distribution network design is about the strategic placement of warehouses, distribution centers, and points of delivery to ensure delivery is cost-effective and timely. This component is particularly complex in global supply chains, where decisions about facility location and transportation links must consider trade-offs between economies of scale and proximity to markets. Logistics engineering applies optimization tools such as linear programming and simulation to evaluate these trade-offs. Procter & Gamble's supply chain transformation, which consolidated distribution centers while maintaining customer service levels, demonstrates how effective network design enhances logistics performance [7].

Today, information and communication technologies are the core of modern logistics systems; they bridge physical flows with digital intelligence. ICT enables the sharing of data, gaining real-time visibility, and making timely decisions using enabling technologies like IoT, blockchain, and cloud computing. Logistics engineers design systems that implement ICT into every component—from tracking shipments in transit to automating inventory replenishment. For example, the collaboration of Maersk and IBM in developing the blockchain-based TradeLens platform demonstrates how ICT can enhance transparency while reducing paperwork and delays in international shipping. However, despite its potential and initial success in attracting major players, TradeLens was ultimately discontinued in late 2022. This decision stemmed from the challenge of achieving full global industry collaboration, a critical factor for the platform's widespread adoption and long-term viability. The TradeLens case serves as a valuable lesson in the complexities of implementing blockchain solutions within established industries, underscoring the importance of broad industry buy-in and the need to address concerns about competition and data ownership [2]. While TradeLens may not have achieved its initial goals, it paved the way for future innovation in the logistics sector. The lessons learned from this ambitious project will undoubtedly inform future efforts to leverage blockchain and other ICT solutions to optimize global trade and supply chain management.

The area of reverse logistics involves products moving backward through the supply chain for return, recycling, or waste management. This area of logistics has come into prominence with increasing electronic commerce and sustainability concerns. Logistics engineering applies network modeling and life cycle analysis to design systems that facilitate efficient product recovery. An example is Dell's recycling program, which uses logistics engineering to manage the return and refurbishment of electronic products, reducing environmental impact while recovering valuable materials [11].

Last, there are human resources and organizational systems, less concrete but not less important. Workforce capabilities in logistics engineering must be developed in accordance with technological and operational objectives through the application of workforce planning, training, and ergonomic design. The collaborative use of robots—known as cobots—further supplements the logistics system, adding value through enhanced productivity by keeping workers safe. For instance, the employment of cobots at Ocado's automated warehouses reflects a fine mixture of human ingenuity with engineering acumen.

It is the integration of these components into one seamless logistics system that is the signature of logistics engineering. Logistics engineers apply tools from operations research, systems analysis, and advanced analytics to develop resilient, efficient systems responsive to market dynamics and customer expectations. Although big steps have been taken, challenges remain due to the sophistication of contemporary global supply chains, the call for sustainability, and fast-paced technological change. Nevertheless, logistics engineering provides the framework and tools needed to address the new challenges and drive innovation across industries.

1.2.3 RELATIONSHIP BETWEEN LOGISTICS AND SUPPLY CHAIN MANAGEMENT

The relationship between logistics and SCM is complex and complementary, with each discipline playing a major role in achieving operational excellence for the modern business. Where logistics focuses on efficiently moving and storing goods, information, and services, the scope of supply chain management encompasses the strategic coordination of all activities within the supply chain network, from the sourcing of raw materials to the final delivery of products to consumers. Integration of the two areas can lead to a synergistic effect, enhancing organizational performance and customer satisfaction.

Logistics is sometimes understood as part of supply chain management, though it highlights some tactical and operational processes comprising transport, warehousing, inventory, and order fulfillment. In this case, the movement of goods from a manufacturing facility to the distribution center will be considered a logistical activity under SCM. It also encompasses strategic issues related to the management of supplier relationships, demand forecasting, production planning, and sustainability concerns. The holistic approach means that all parts of the supply chain are aligned to accomplish goals with a view to maximizing value creation for all stakeholders. Thus, effective integration across the supply chain is essential [7].

The interdependence between logistics and SCM can be well understood in practical application from companies such as Apple. The efficiency of Apple's supply chain has been a model, in which logistics has played the most central role. Advanced logistical strategies are employed, such as just-in-time inventory and optimized transport routes, to reduce lead times and lower costs. Meanwhile, the management of its supply chain—robust supplier relationships and tight demand forecasting, among other tactics—keeps its logistical functions well-served and integrated with the broader aims of the company.

Inventory management is one of the major areas where logistics and SCM overlap. Inventory management is a delicate balance between stocking enough to satisfy customer demand and minimizing excess inventory to reduce holding costs. This balance is well exemplified by Walmart through the use of logistics solutions such as cross-docking, where goods are transferred directly from inbound to outbound transportation without intermediate storage. It enables this logistical innovation because the integrated real-time data analytics in the advanced supply chain management of Wal Mart facilitate forecasts of demand patterns with the associated inventory levels accordingly [27].

The result of technological changes has only made the connection between logistics and SCM stronger. Digital solutions such as ERP and TMS make communication and coordination among different supply chain functions possible. For instance, Amazon's supply chain is greatly dependent on its proprietary logistics and inventory management software, whereby all the warehouse operations, transport, and customer orders are integrated into a single system. This integration not only smoothes the logistical processes but also enables Amazon to attain unparalleled levels of efficiency and responsiveness in its supply chain, as Waller and Fawcett [28] assert.

Increasingly, the focus on sustainability has underlined the interconnection between logistics and SCM. For instance, sustainable supply chain practices require logistics systems that are not harmful to the environment, thus, minimizing carbon emissions from transport or using energy-efficient warehouses. For instance, DHL has created its GoGreen program, which inlines under the general SCM efforts sustainable logistics solutions, such as route optimization and electric vehicles, to reach carbon neutrality across the supply chain. In this respect, logistics may be referred to as a starting point in reaching supply chain sustainability goals.

Considering the cases of synergy between the two concepts, logistics and SCM, there are still some issues. One significant issue is the misalignment between logistics operations and supply chain strategies. For example, a company may invest heavily in reducing transportation costs through logistics optimization but fail to align these efforts with broader supply chain objectives, such as improving customer service levels or enhancing supply chain agility. Such misalignment can lead to suboptimal outcomes and undermine the overall effectiveness of the supply chain [7].

Other challenges also include the increase in complexity within global supply chains. Logistics operations have to make their way through a myriad of variables, fluctuating fuel prices, regulatory compliance, and geopolitical risks. Effective SCM frameworks are needed for coordinating these variables and ensuring that logistics activities are resilient and adaptive to external disruptions. The COVID-19 pandemic accentuated this problem, as the companies faced some unprecedented supply chain disruptions that tested the need to integrate logistics more with SCM. Companies that invested in digital supply chain solutions and flexible logistics networks coped better in the crisis and adapted fast to the fast-changing market conditions.

1.2.4 SYSTEMS APPROACH TO LOGISTICS

A systems approach to logistics is a conceptual framework that looks at logistics in terms of an interacting and interdependent network of subsystems or components working in collaboration toward the broader organizational objectives. It underlines the interrelationships among the various components, like transportation, warehousing, inventory management, and information systems, pointing out that the performance of the system as a whole is usually greater than the sum of its individual parts. This perspective lets organizations comprehend the modern complexities of supply chains while seeking ways to enhance efficiency, reliability, and adaptability.

The systems approach is derived from systems theory, where the behavior of a system emerges from interaction among its constituent parts. In logistics, this will mean that the approach must be integrated, with processes and functions coordinated to minimize duplication, reduce costs, and improve service quality. For instance, a manufacturing company using the systems approach in its operations will schedule production to match current inventory levels and available transport, hence eliminating wastages and loss of time [7].

One well-known application of the systems approach in the field of logistics is supply chain network design. In this respect, traditional methods normally consider transportation, warehousing, and inventory as separate entities while the systems approach embraces them as related parts of an integrated process. For example, reducing inventory levels could reduce the cost of storage but may be associated with increased frequency of transportation, hence driving overall costs up. A systems approach empowers firms to identify an optimal balance by modelling and simulating such interdependencies. For example, Coca-Cola uses sophisticated simulation tools for designing its supply chain network where the company has to position its manufacturing plants, distribution facilities, and retail points to meet demand at an optimal cost [9].

Another important ability of the systems approach is how it can handle dynamics regarding the nature of the logistics systems. Supply chains operate in environments characterized by uncertainty, including fluctuating customer demand, geopolitical instability, and climate-related disruptions. The systems approach embeds flexibility and resilience into the design of logistics that enable organizations to adapt to such changes. For instance, during the COVID-19 pandemic, many companies experienced huge disruptions to their supply chains. Companies that take the systems approach include Procter & Gamble, who harnessed leading-edge analytics and digital technology to redesign the supply chains in these unprecedented times of humanitarian crises [16]. Besides strategic planning, the system's approach impacts even on regular logistics operation processes. Take as an illustration the incorporation of TMS within WMS. By linking these platforms, organizations can achieve greater visibility into their logistics operations, enabling dynamic routing, efficient load planning, and real-time tracking. Amazon exemplifies this integration, using sophisticated algorithms to coordinate last-mile delivery, inventory placement, and order fulfillment within its extensive logistics network. This systemic coordination allows Amazon to meet stringent delivery timelines while managing costs effectively [28].

Another area where the systems approach has proved invaluable is in sustainability. Logistics systems are one of the major contributors to environmental degradation, especially through greenhouse gas emissions from transportation and energy-intensive warehousing. A systems perspective allows organizations to identify synergies between economic and environmental objectives. For instance, Unilever uses a systems approach to minimize carbon emissions in its supply chain by incorporating renewable energy sources, route optimization, and engaging suppliers in the introduction of sustainable practices. This not only helps Unilever achieve its corporate social responsibility objectives but also minimizes operational costs.

Despite the advantages, the adoption of a systems approach to logistics is not without its challenges. Key barriers include the fact that, in essence, large-scale system modeling and analysis is a complex process. Most logistics systems are bound to include a multitude of stakeholders, often conflicting objectives, and large amounts of data, making comprehensive modeling rather challenging. AI and machine learning technologies have so far, to a certain extent, alleviated some of these challenges by enabling the analysis of complex datasets and predicting system behavior for a variety of different scenarios. Walmart's use of predictive analytics to forecast demand and optimize inventory levels is one example of the potential of technology in enhancing the systems approach.

The second challenge involves collaboration along the supply chain. The systems approach requires effective communication and sharing of data among suppliers, manufacturers, distributors, and retailers. However, this is often hindered by differences in technological capabilities, cultural barriers, and competitive interests. This is where blockchain technology came in as a solution to this problem, offering a secure and transparent platform for information sharing.

1.2.5 DESIGN AND OPTIMIZATION OF LOGISTICS SYSTEMS

The Figure 1.3 below captures the interaction of inputs, factors, constraints, and outcomes involved in the design and optimization of logistics systems by putting into perspective how these elements interact to realize efficient and resilient supply chain operations. Inputs such as data, resources, and technology are foundational to drive decision-making. Data includes a broad spectrum of critical information including demand forecasts, cost of transportation, and supplier performance metrics. For instance, Walmart applies predictive analytics in decoding historical sales data to set up inventory at levels that allow it to efficiently meet customer demand. Resources in designing logistics also involve such key things as transportation fleets, warehouses, and skilled people. Amazon fulfillment centers, which are strategically positioned based on their proximity to demand, help illustrate how these resources can be optimized to facilitate more service efficiency. Technology further enhances this process: Advanced tools, such as Transportation Management Systems and the Internet of Things devices, allow real-time tracking, automation, and proactively solve problems. DHL is a good example of IoT-enabled sensors that track temperature-sensitive goods, like vaccines, while they are on the move.

As these inputs are applied, they influence several critical factors, which in turn guide the optimization process. Efficiency is often the first consideration because organizations try to cut costs while maintaining high levels of performance. FedEx shows this focus through its sophisticated route optimization algorithms that reduce delivery times and fuel consumption. Another key factor is sustainability, which has

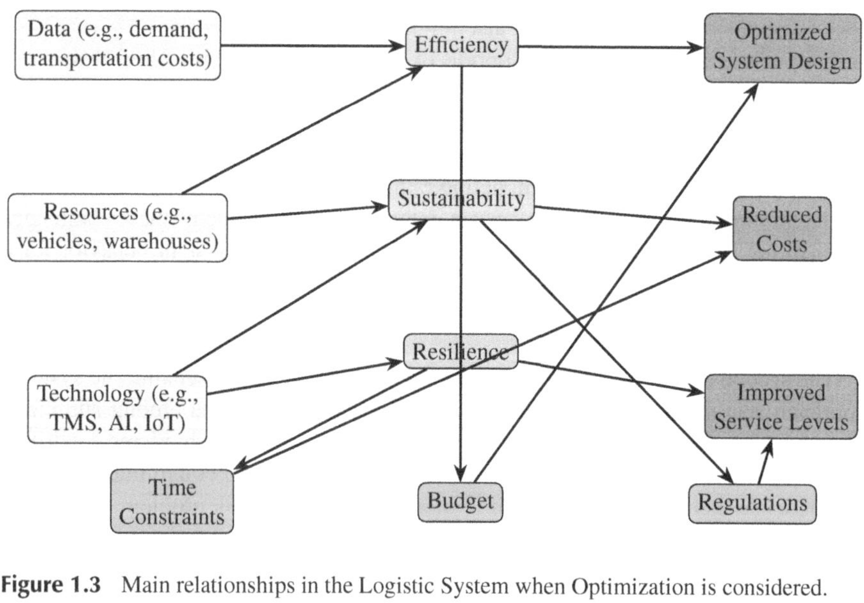

Figure 1.3 Main relationships in the Logistic System when Optimization is considered.

become increasingly significant as companies address environmental concerns. For instance, Unilever uses electric vehicles in its logistics, which reduces carbon dioxide emissions and makes the practice of logistics more sustainable in line with the world's goals of sustainability. Resilience is another key factor, especially in today's turbulent markets. Digital twins enable Procter & Gamble to simulate any supply chain disruption and make necessary arrangements in advance, like those that occurred during the COVID-19 pandemic.

In spite of these objectives, there are certain constraints within which logistics systems have to function: budgetary limits, time pressures, and regulatory compliance. Budgetary limits often put a damper on the scale of solutions, as small businesses may well focus on regional distribution hubs rather than expansive global networks. Time pressures are especially severe in industries where customer expectations for rapid delivery are high. Companies like DoorDash, working in the highly competitive food delivery business, have invented last-mile strategies to ensure timely fulfillment. Compliance with regulations adds another layer of complexity, especially in global supply chains. That's where the blockchain-based TradeLens platform created by Maersk comes in: enabling a secure and transparent way to manage the documentation of trade and adhere to international regulations.

The outcomes of good logistics system design are numerous and quite tangible. The optimized system design results in efficient layouts for warehouses, well-coordinated transportation networks, and streamlined workflows that altogether improve the functionality of the entire system. A typical example is the utilization of advanced simulations by Coca-Cola in planning its global distribution network with the view of minimizing costs while ensuring product availability. Other critical outcomes are cost reductions, whereby the implementation of Just-in-Time logistics by Toyota remains a benchmark to minimize inventory carrying cost and improve operational flexibility. Finally, improved service levels remain important since customer satisfaction will be directly linked to the reliability and speed of its logistics operations. UPS has already shown this through its ORION system, which uses data analytics to shave precious minutes off its delivery times, boosting efficiency while slashing emissions.

The diagram puts together that logistics systems are a dynamic interplay of elements: inputs fuel the process, factors guide it, constraints shape its boundaries, and outcomes reflect success. In short, the case of Apple will be used to powerfully illustrate this holistic approach. By leveraging demand data, a robust network of suppliers, and state-of-the-art tracking technology, Apple has developed a logistics system that creates efficiency, reduces packaging waste through sustainability, and enhances the resilience of its operations via supplier diversification. These are shaped by international tariffs and shipping delays, yet the outcomes are seen in the capability to deliver products across the world to customers with very high satisfaction. This systems-oriented perspective underlines how important it is to align all the elements coherently in order for organizations to turn logistical challenges into opportunities for innovation and competitive advantage.

1.2.6 NETWORK DESIGN AND ROUTING

Network design and routing problems in logistics engineering represent a critical area of study that seeks to optimize the configuration and operation of supply chain networks. These problems are underpinned by a combination of mathematical modeling, operations research, and systems theory, with the overarching goal of minimizing costs, improving efficiency, and meeting service level requirements. The theoretical basis for these problems usually draws from graph theory, combinatorial optimization, and linear programming.

It is essentially about finding the best placement of facilities such as warehouses, distribution centers, and manufacturing plants in a supply chain. It should result in a structure that minimizes total costs of transportation, inventory holding, and facility operating while being able to meet customer demand. The capacitated facility location problem is the classical formulation in this context, where decisions on adding constraints, such as facility capacity and fixed operating costs, are integrated into the decision-making process. An example of a network design would be how Walmart conducts its operations of placing distribution centers to minimize the distance to its stores, aiming at lowering transportation costs and enabling timely replenishment of inventory.

Although the basic routing problem concerns the development of most economic routes for efficient carriage of shipments in the whole network, Vehicle Routing Problem from a more general theoretical setting provides the way to discover an optimal number of routes taken up by vehicles. Some of the variants developed on VRP include the Capacitated VRP, Time-Window VRP, and Multi-Depot VRP. Each variant has been developed to fulfill specific logistical requirements. For example, lower-bound constraints on vehicle capacity are part of a CVRP, while in VRPTW, constraints on delivery time windows ensure that customers receive goods within certain specified periods.

The UPS implementation of the ORION system is one practical application of solving routing problems. ORION uses sophisticated algorithms to solve a variant of the VRP, considering factors such as traffic conditions, package priorities, and driver schedules. This has saved the company millions of gallons of fuel annually, greatly reduced emissions, and greatly improved delivery efficiency by dynamically adjusting the delivery routes.

The network design and routing problems are not only cost minimization problems but also a trade-off between the different factors inherent in supply chain operations. A well-designed network reduces lead times by bringing facilities closer to demand centers, while this may increase operating costs because of additional facilities. Similarly, the routes must be optimized for transportation costs without sacrificing service quality, since very long routes or excessive delays result in dissatisfied customers. For instance, in e-commerce logistics, there is a pretty explicit trade-off between the cost of operating an urban fulfillment center and the service benefit of same-day delivery.

The difficulty in solving these problems increases with the complexity of the network. Global supply chains, typified by many stakeholders, cross-border regulations, and volatile demand, require advanced computational approaches. Genetic algorithms, simulated annealing, and ant colony optimization are some of the heuristic and metaheuristic algorithms widely used to solve large-scale network design and routing problems. These methods provide near-optimal solutions within reasonable computational times; hence, they can be effectively used in real-world applications. For example, DHL uses metaheuristic algorithms in order to optimize its international shipping routes so that goods can move efficiently across its global network while considering local regulations and constraints.

In the next chapters you will learn tools helping to build models and to carry out analyses. In addition, eEmerging technologies have also given new dimensions to network design and routing problems. Artificial intelligence and machine learning now allow for predictive modeling of demand patterns and traffic conditions, furthering the adaptability of logistical networks. The Internet of Things adds to real-time visibility, enabling dynamic routing adjustments based on current conditions. Blockchain technology has brought much more transparency and trust among stakeholders in the supply chain, especially in industries such as pharmaceuticals and food, where traceability is of prime importance.

One of the key challenges in network design and routing is resilience. Supply chains are severely disrupted by catastrophes, pandemics, or geopolitical tensions. For example, Procter & Gamble uses digital twins-a virtual representation of their supply chain network-to simulate disruptions and build contingency plans. These simulations provide insight into decisions about alternative routes or temporary facility setups that will keep products flowing to consumers despite external shocks.

1.3 MATHEMATICAL AND ANALYTICAL TOOLS IN LOGISTICS

Mathematical and analytical tool support in logistic processes are key aspects for developing and optimizing and handling logistics systems. These have been primarily supported by operations research, simulation, and modeling techniques to show clear-cut insight into the decision maker for complex problems associated with various facets of the logistical systems. These techniques are not just conceptual; they are real-life, indispensable tools for finding solutions to real-life logistics problems: from network optimization and inventory management to dynamic resource allocation under conditions of uncertainty.

Operation research has long been a cornerstone in logistics engineering. Drawing its roots from mathematical optimization, probability theory, and statistical analysis, it gives a formalized approach to decision-making. Operations research allows the formulation of logistics challenges as mathematical problems, whose solutions can be found to be optimal or near-optimal. The most common techniques employed in solving problems including transportation routing, warehousing location, and production scheduling involve linear programming, integer programming, and dynamic programming. As an example, one could think of the classic application of linear programming to the so-called transportation problem that consisted of minimizing the cost of shipment from several origins to several destinations in a way that demand

and supply constraints are met. The details of these methods, including their practical applications in modern supply chains, follow in the chapters ahead.

Simulation and modeling are perhaps the most valued techniques in logistics, since complex systems-amenable analytical solutions hardly exist-can be represented and analyzed. In this respect, the tools will allow practitioners to create digital twins of logistic networks and comprehend their performances in different scenarios. Simulation models can have embedded stochastic elements, such as fluctuating demand or uncertain lead times, to show insights into system behavior under uncertainty. Examples include the frequent use of Monte Carlo simulations in inventory management to identify the amount of safety stock required to minimize stockout risk while keeping carrying costs in check. Another powerful technique is discrete-event simulation, which is widely used in both warehousing and transportation systems to optimize workflows and resource utilization. These methods not only enable problem-solving but also strengthen strategic planning by enabling the stakeholders to try out various decisions in a no-risk environment.

The most promising area of application of simulation and modeling is automatic control theory, a subject that will be given special emphasis in the chapters to come. Control theory, conventionally rooted in engineering fields like robotics and aerospace, has been gaining increasing relevance in logistics systems. The core of control theory is the concern with maintaining desired system behaviors through feedback mechanisms. This implies, in logistics, dynamically adjusting variables such as inventory levels, production rates, and transportation schedules in response to real-time data. For example, the warehouse management systems installed with feedback mechanisms will reorder inventory whenever the level of inventory has gone below a set threshold level, all this happening in an uninterrupted manner. In like vein, fleets of vehicles that are autonomous will leverage control theory in routing decisions based on the current traffic and priority of delivery.

The embedding of automatic control theory in logistics brings many advantages. It sets the basis for developing adaptive systems that are prompt in responding to any kind of disruption, say sudden surges in demand or disruptions in supply chains. Such adaptiveness is very critical in global supply chains where variability and uncertainty are intrinsic. Control theory allows the development of predictive systems that anticipate future states based on current and historical data, further enhancing the resilience and efficiency of logistics operations. For instance, predictive maintenance systems in transportation logistics, guided by control principles, can forecast equipment failures and schedule repairs proactively, reducing downtime and operational costs.

In addition, the chapters discuss how control theory interacts with other analytical tools-such as machine learning and optimization algorithms-in the development of intelligent logistics. Combining predictive analytics with control theory allows companies to obtain a level of foresight and adaptability they could not otherwise achieve. A good example might be dynamic pricing strategies in logistics, where prices for transportation may be adjusted instantly based on a given capacity utilization and/or market demand.

The succeeding chapters will attempt to provide a holistic understanding of how these mathematical and analytical tools have enabled the logistics professional to design systems that are not only efficient but also robust and adaptive. The inclusion of control theory as a central theme underlines its critical role in modern logistics, filling the gap between static optimization models and dynamic real-world operations. By emphasizing both the theoretical bases and practical uses of these tools, this book prepares readers to effectively respond to complex issues arising in logistics systems in an ever-dynamic and connected world.

1.4 TECHNOLOGICAL ADVANCEMENTS IN LOGISTICS

The integration of modern technologies like automation and robotics, IoT, blockchain, and AI has really transformed the way goods are stored, transported, and delivered. This book will explain each of these advancements in detail and provide readers with an overview of their application, benefits, and challenges t hat may arise in logistics.

Automation and robotics within the warehouse are cornerstones of technological progress in logistics that unmistakably contribute to increasing the operational efficiency and precision of an enterprise. The innovations overcome traditional bottlenecks, such as manual handling delays, inventory inaccuracies, and labor shortages. This chapter discusses how automated guided vehicles, robotic picking systems, and high-density storage solutions have been implemented in order to automate fulfillment centers by companies such as Amazon and Ocado. It will also discuss implications for workforce dynamics due to these technologies and necessary shifts in labor skills, striking a balance in views of opportunities and challenges thrown up by these technologies.

IoT has turned out to be a game-changing enabler in smart logistics systems. This technology thus provides the capability for real-time monitoring and information exchange between nodes of the supply chain-a new dimension toward giving unparalleled visibility to the respective logistics operations. Examples will be highlighted, such as IoT-enabled sensors used in cold chain logistics for temperature and humidity monitoring of sensitive products like vaccines. It will also look at how IoT devices contribute to predictive maintenance for vehicles and equipment, reducing downtime and operational disruptions. By interweaving theoretical insights with real-world applications, the chapter will emphasize the critical role of IoT in building agile and responsive logistics systems.

Blockchain technology has gained considerable attention for its potential to improve supply chain transparency and trust among stakeholders. Chapter eight will explore how blockchain creates tamper-proof records of transactions in a secure way, ensuring accountability and traceability across the supply chain. The chapter will also discuss other challenges, such as scalability and regulatory compliance, in a balanced manner to provide a realistic view of how blockchain can change the face of global logistics. The chapter will explore AI applications such as demand forecasting, route optimization, and warehouse layout planning that help companies reduce costs and improve service quality. Case studies, such as UPS, which relies on

AI-powered systems to help optimize delivery routes, will illustrate the pragmatic and practical benefits of this technology. The chapter also covers some of the key ethical discussions of AI in logistics, including issues around algorithmic bias and data privacy.

As indicated, chapter eight will look at these technological developments and provides an overall analysis of how they interlink to drive innovation and competitiveness in the logistics sector. It will, therefore, be a source book for academics, practitioners, and decision-makers who try to make sense of the complexity of modern logistics in a world increasingly driven by technology. The chapter will consequently strive to bridge the gap between theoretical understanding and practical application, hence providing valuable insights into the opportunities and challenges associated with the integration of these technologies into logistics systems.

1.5 EXERCICES AND HOMEWORK

1.5.1 CASES

- **Operations Research in Logistics**
  - A company operates two large warehouses and delivers goods to three major customer regions. The company is facing challenges in balancing transportation costs with meeting customer demand efficiently. Describe how the company could approach this problem using operations research methods without delving into mathematical models. Discuss the types of data required, possible strategies (e.g., consolidating shipments, prioritizing routes), and how these might impact cost and service levels.
  - Consider a city where a logistics provider is setting up a delivery network to serve a growing urban population. Traffic congestion is a major concern, and the company wants to identify the shortest paths for its delivery vehicles while minimizing fuel consumption. Explain the key factors they should consider when designing the network and how techniques from operations research can help. Include examples of how different strategies might be applied in practice.
  - A company has a fleet of trucks that must deliver goods to customers within strict time windows. However, delays due to weather or traffic often disrupt schedules. Describe how sensitivity analysis could help the company better understand these disruptions and make more informed decisions about resource allocation and route planning.
- **Simulation and Modeling Techniques**
  - A distribution center handles hundreds of orders daily, but customer complaints about delays are increasing. The manager is considering a simulation study to identify bottlenecks in the order processing system. Describe how a simulation model could be used in this scenario, focusing on the steps involved in building the model (e.g., defining processes, collecting data, testing scenarios). Discuss the potential outcomes of the simulation and their implications for improving operations.

- A retail company is deciding how much safety stock to maintain for its best-selling products. Customer demand varies seasonally, and the company wants to avoid both stockouts and overstocking. Without using equations, explain how a Monte Carlo simulation could help the company make better inventory decisions. Include a discussion of what kind of data they need to gather and the types of scenarios they could test.
- A manager of a warehouse is considering changes to the replenishment strategy to reduce operational costs. The warehouse follows a fixed replenishment schedule, but demand is highly variable, leading to frequent overstocking or stockouts. Describe how a simple simulation experiment could help them test different inventory policies and evaluate their impact on cost and service levels.

1.5.2 EXPECTED DELIVERABLES

Students are expected to submit written analyses for each problem, emphasizing qualitative insights, real-world implications, and thoughtful discussion of trade-offs and potential solutions, avoiding unsing LLM technology. To this end, enough detailed context and specific reasoning is expected, while expected perpexity should be medium and high. Diagrams or flowcharts may be included where appropriate to enhance explanations.

1.5.3 ASSESSMENT CRITERIA

1. **Understanding:** Demonstrates a clear grasp of logistics concepts and their application to real-world problems.
2. **Clarity:** Clear explanations of methodologies, assumptions, and conclusions.
3. **Critical Thinking:** Depth of analysis, including insights into trade-offs, constraints, and implications of the proposed solutions.
4. **Presentation:** Professional quality of visualizations, graphs, and reports.
5. **Creativity:** Suggests innovative ideas or approaches to address logistics challenges.

Section II

Mathematical Models in Logistics

2 Mathematical Models in Logistics

2.1 INTRODUCTION

Modern logistics lies at the heart of global commerce and economic growth, serving as the intricate network that moves goods, services, and information from their points of origin to end-users. As markets become more interconnected and customer expectations continue to rise, logistics processes have grown ever more complex. They must deal with long supply chains spanning multiple continents, volatile demand patterns, diverse transportation modes, and stringent service-level requirements.

Mathematical modeling provides a framework for understanding, predicting, and optimizing these logistics operations. Among the most powerful tools for capturing time-dependent processes are *differential equations*, which describe how critical variables—such as inventory levels, production rates, or the occupancy of shipping channels—change over time in response to internal and external factors. In addition, *Laplace transforms* enable the transformation of these time-domain equations into a different realm (the s-domain), where complex dynamics can often be analyzed and solved more directly. By uniting these mathematical techniques with an understanding of logistics processes, engineers and managers can make more informed decisions, design robust control strategies, and respond effectively to uncertainties in the supply chain.

In this chapter, we introduce the mathematical foundations of differential equations and Laplace transforms, focusing on their application to logistics systems. We will treat variables such as inventory or flow rates as functions of time and model them using differential equations that capture both continuous inflows and outflows. We will then illustrate how Laplace transforms simplify the process of solving these equations, particularly when dealing with piecewise or impulse-like demand patterns, lead times, or reordering policies. The overarching theme is to see how all these elements—inventory flows, response times, and distribution routes—interact within a larger logistics network and how mathematical models can provide valuable insights.

2.1.1 WHY MATHEMATICAL MODELING MATTERS IN LOGISTICS

Effective logistics operations depend on accurately anticipating or responding to changes in supply, demand, and transportation conditions. Overly simplistic models can lead to poor service quality (e.g., stockouts or late deliveries) or inflated costs (e.g., excess inventory or emergency shipments). Mathematical modeling, especially when it harnesses the power of differential equations, offers several key benefits:

DOI: 10.1201/9781003627906-2

1. **Predictive Capability:** By calibrating differential equation models with historical data, it becomes possible to forecast how inventory levels or transportation queues will evolve, helping planners detect potential bottlenecks in advance.
2. **Systematic Optimization:** Once the relationships within a logistics system are encoded in mathematical terms, they can be manipulated and optimized subject to constraints such as capacity, budget, or time. This fosters data-driven decision-making, replacing guesswork with actionable strategies.
3. **Interconnected Systems View:** In modern supply chains, small local changes can have significant ripple effects across other processes. Mathematical models help visualize these interdependencies clearly, revealing how one decision may propagate through upstream and downstream nodes.
4. **Risk Management and Robustness:** Logistics networks are prone to uncertainties, whether from demand fluctuations, transit delays, or machine breakdowns. Modeling these variations enables scenario analysis and leads to robust designs that maintain acceptable performance under stress.

Because logistics systems often operate in a dynamic environment, a framework based on differential equations is especially relevant: it captures how certain key variables increase or decrease continuously over time. From these equations, we also learn how the system responds to external inputs (like demand spikes) and internal feedback mechanisms (like reorder policies).

2.1.2 DIFFERENTIAL EQUATIONS IN LOGISTICS AND INVENTORY MANAGEMENT

At its core, a *differential equation* ties the rate of change of a variable to other variables or to the variable itself. In logistics and supply chain management, some of the most common variables include:

1. $I(t)$: Inventory level at a warehouse or distribution center.
2. $D(t)$: Demand rate (units per time).
3. $Q(t)$: Production or replenishment rate (units per time).
4. $T(t)$: Transit or throughput capacity in transportation links.

An elementary example arises when modeling the inventory $I(t)$ in a single facility:

$$\frac{dI(t)}{dt} = Q(t) - D(t).$$

Here, the inventory changes at a rate determined by the difference between what flows in ($Q(t)$) and what flows out ($D(t)$). Although this model might appear simplistic, it forms the backbone for more sophisticated equations that incorporate:

1. **Reorder Policies and Lead Times:** When $I(t)$ dips below a threshold, a new order is placed, but it arrives after a certain lead time. Such time delays can be modeled with shifted demand or supply terms.

2. **Capacity Constraints:** Production or replenishment rates may be capped, or shipments could be limited by truck capacities. Nonlinear or piecewise definitions of $Q(t)$ might be needed.
3. **Forecast Adjustments:** Demand $D(t)$ may fluctuate based on trends, seasonal patterns, or real-time data. In some cases, $D(t)$ can be random, motivating extensions to stochastic differential equations.

In more extensive logistics networks, each facility might have its own inventory function $I_i(t)$ and a corresponding set of inflows and outflows. The network becomes a system of coupled differential equations:

$$\frac{dI_i(t)}{dt} = \sum_{j \in \mathscr{P}(i)} f_{ji}(I_j(t)) - \sum_{k \in \mathscr{C}(i)} g_{ik}(I_i(t)),$$

where $\mathscr{P}(i)$ denotes the set of predecessor nodes (suppliers) to node i, and $\mathscr{C}(i)$ denotes the set of customer or successor nodes. The functions f_{ji} and g_{ik} might incorporate production rates, capacity constraints, and other complexities.

By solving or simulating these equations, we gain insight into whether the logistics network can handle variations in demand without suffering large oscillations or shortages, how delays propagate through the system, and which parameters are most influential in stabilizing or destabilizing the network.

2.1.3 RESPONSE TIMES AND SYSTEM DYNAMICS IN LOGISTICS

Response time in a logistics context refers to how quickly a supply chain or logistics system can adjust to variations in demand, disruptions in supply, or other unexpected events. Typical questions include:

1. How fast can a warehouse restock once inventory falls below a safety threshold?
2. How rapidly can a distribution system reroute shipments when a transportation corridor is closed?
3. What is the system's transient behavior when a sudden demand spike occurs?

In differential equation terminology, these responses often manifest as exponential decays or oscillations. For a simplified single-node inventory system, one might write:

$$\frac{dI(t)}{dt} = -\alpha\big(I(t) - I_{\text{target}}\big),$$

where $\alpha > 0$ is a responsiveness parameter controlling how aggressively we try to revert inventory to the target level I_{target}.

Derivation of the solution. We can rewrite the equation in standard linear form:

$$\frac{dI}{dt} + \alpha I(t) = \alpha I_{\text{target}}.$$

An integrating factor is

$$\mu(t) = e^{\int \alpha\, dt} = e^{\alpha t}.$$

Multiplying through by $\mu(t)$ gives

$$e^{\alpha t}\frac{dI}{dt} + \alpha e^{\alpha t} I(t) = \alpha I_{\text{target}}\, e^{\alpha t} \quad \Longrightarrow \quad \frac{d}{dt}\left(e^{\alpha t} I(t)\right) = \alpha I_{\text{target}}\, e^{\alpha t}.$$

Integrating both sides with respect to t,

$$e^{\alpha t} I(t) = \alpha I_{\text{target}} \int_0^t e^{\alpha s}\, ds + C = \alpha I_{\text{target}} \left[\frac{e^{\alpha s}}{\alpha}\right]_{s=0}^{s=t} + C = I_{\text{target}}\left(e^{\alpha t} - 1\right) + C.$$

Solving for $I(t)$,

$$I(t) = I_{\text{target}}\left(1 - e^{-\alpha t}\right) + Ce^{-\alpha t}.$$

The constant C is determined from the initial condition $I(0) = I_0$:

$$I(0) = I_{\text{target}}(1-1) + C = C \quad \Longrightarrow \quad C = I_0 - I_{\text{target}}.$$

Hence, the explicit solution is

$$\boxed{I(t) \;=\; I_{\text{target}} \;+\; \left(I_0 - I_{\text{target}}\right) e^{-\alpha t},}$$

which shows that $I(t)$ exponentially approaches the target level with time constant

$$\tau = \frac{1}{\alpha}.$$

Larger α means a faster correction to the desired level, which may reduce shortages but also risk larger overshoots or oscillations if the system is more complex.

In multi-echelon supply chains, local response parameters at individual warehouses or distribution centers can produce system-wide effects. If each node attempts to correct its inventory too aggressively, one might witness the "bullwhip effect," where small demand variations at the retail level amplify into large swings upstream. Studying the equations that govern these interactions is critical for designing control policies that maintain stable operation while meeting service-level requirements.

2.1.4 LAPLACE TRANSFORMS FOR LOGISTICS SYSTEMS

While differential equations provide the language for describing logistics dynamics, they can become cumbersome to solve directly in the time domain. *Laplace transforms* offer a powerful approach to simplify the mathematics and glean deeper insights into how the system behaves under different input signals.

The Laplace transform of a function $f(t)$, defined for $t \geq 0$, is

$$F(s) = \mathscr{L}\{f(t)\} = \int_0^\infty e^{-st} f(t)\, dt,$$

where s is a complex variable. This transform turns calculus in the time domain into algebra in the s-domain, with several properties particularly useful in logistics:

1. **Transforms of Derivatives:**
$$\mathscr{L}\left\{\frac{df(t)}{dt}\right\} = sF(s) - f(0).$$
This property converts a differential equation into an algebraic equation in terms of s.
2. **Linearity:**
$$\mathscr{L}\{af(t) + bg(t)\} = aF(s) + bG(s),$$
which allows superimposing multiple effects easily.
3. **Convolution Theorem:**
$$\mathscr{L}\{f(t) * g(t)\} = F(s)G(s),$$
where $*$ denotes convolution in time. Convolution naturally arises in logistics systems when the outflow from one node becomes inflow for another after a delay or filtering process.
4. **Initial and Final Value Theorems:** These theorems help determine $f(0)$ and $\lim_{t\to\infty} f(t)$ from the behavior of $F(s)$ as s goes to ∞ or 0.

Consider the simple inventory model
$$\frac{dI(t)}{dt} + \alpha I(t) = R(t),$$
where $\alpha > 0$ models an outflow rate proportional to $I(t)$ (or a decay toward a desired setpoint) and $R(t)$ represents replenishment or demand signals. Taking the Laplace transform and using the above properties gives:
$$sI(s) - I(0) + \alpha I(s) = R(s),$$
which simplifies to
$$I(s) = \frac{I(0) + R(s)}{s + \alpha}.$$
Hence, solving for $I(t)$ in the time domain requires finding the inverse Laplace transform of this quotient. When $R(t)$ is a piecewise constant or a step function (common in logistics where demand changes at known times), inverse transforms are often easier to handle than solving the original differential equation directly.

Moreover, in more complex networks with multiple interacting inventories, the Laplace transform offers a systematic way to examine each node's transfer function. Poles and zeros of these transfer functions indicate potential oscillatory behaviors or sensitivities to certain frequencies of input demand. Armed with this information, logistics planners can identify design parameters (e.g., reorder points, lead times, and production capacity) that promote stability and mitigate large swings in inventory levels.

2.1.5 MODELING INVENTORY AND FLOWS AS SIGNALS

A productive viewpoint in modern logistics is to treat inventory, demand, and shipment quantities as *signals* in a system-theoretic sense. In this signal-based perspective:

1. **Input Signals:** Demands or orders $D(t)$, which may be deterministic (scheduled) or stochastic (random arrivals).
2. **State Signals:** Inventory levels $I(t)$ at different locations, governed by internal accumulation and depletion.
3. **Output Signals:** Shipping or production quantities $Q(t)$ that aim to match or counteract the demands.

Each node or facility in the logistics network can be modeled as a "system" that transforms input signals into output signals. Some transformation rules might be linear, while others might be nonlinear (e.g., saturations due to capacity limits). From a control-theoretic standpoint, we can then design feedback mechanisms that automatically adjust flows in response to measured inventory discrepancies, akin to how a thermostat adjusts heating based on current temperature versus a setpoint.

By analyzing these signals in the frequency domain (via the Laplace or Fourier transform), it becomes clearer whether certain demand patterns might cause amplified fluctuations. If the system has natural oscillatory modes or high-gain feedback loops, even small periodic changes in demand can lead to large inventory swings over time. This phenomenon is especially salient in large supply chains, where *time delays* due to shipping and production lead times can exacerbate instability.

2.1.6 INTERCONNECTED SYSTEMS AND THE LARGER LOGISTICS PICTURE

Few logistics networks operate in isolation; they are often part of broader supply chains that integrate multiple tiers of suppliers, distribution channels, and end consumers. This complexity means that no single equation can capture every detail. Instead, we rely on *system-level* models composed of interconnected nodes. Each node might have its own differential equation describing local processes (e.g., production, storage, or cross-docking), while the arcs between nodes encapsulate transportation delays, capacity constraints, or cost functions.

The ability to link differential equations across various nodes lets planners and engineers analyze the entire flow of goods from raw materials to finished products. By tracking how inventory, demand, and cost evolve over time, one can identify:

1. **Bottlenecks:** Where capacity limits cause backlogs that ripple through the system.

2. **Critical Lead Times:** Which transit or production steps dominate the overall system response time.
3. **Optimal Inventory Placement:** How to strategically position buffers of inventory to protect against demand variability or transport disruptions.

Ultimately, these interconnected models serve as the basis for higher-level tools such as *model predictive control* (MPC) or *real-time optimization*, where decisions (like production scheduling or vehicle dispatching) are updated dynamically based on current system states and forecasts. The Laplace transform remains relevant even in these advanced frameworks, as it provides a concise representation of system dynamics and a means to assess stability under feedback control.

2.1.7 SCOPE OF THE CHAPTER AND LEARNING OUTCOMES

This chapter focuses on providing a clear introduction to differential equations and Laplace transforms within the context of logistics. While we do not shy away from the mathematical rigor, our core aim is to emphasize relevance and applicability. Specifically, we seek to equip you with the following competencies:

1. **Formulating Logistics Problems in the Language of Differential Equations:** Understand how to represent key variables (inventory, demand, capacities) as time-dependent functions and how to write down governing equations that capture inflows and outflows.
2. **Applying Laplace Transforms:** Gain fluency in transforming time-domain models to the s-domain, and learn to use properties of the Laplace transform (e.g., derivative transforms, convolution theorem) to solve practical logistics problems.
3. **Interpreting Solutions and Stability:** Develop the skills to interpret inverse Laplace transforms, identify transient and steady-state behaviors, and diagnose potential instabilities (such as oscillations or bullwhip effects).
4. **Designing Feedback and Control Strategies:** Appreciate how feedback laws—such as proportional control of reorder points—can be analyzed using transfer functions, poles, and zeros to ensure stable and responsive logistics networks.
5. **Synthesizing Concepts for Larger Networks:** Understand the system-level implications of coupling multiple nodes and see how the mathematics scales when dealing with real-world supply chains or distribution systems.

These foundations position you to engage with more advanced logistics topics, such as stochastic modeling (where random shocks play a significant role), discrete-event simulation (relevant when dealing with discrete shipments and scheduling constraints), and hybrid or agent-based approaches (which capture more intricate dynamics and decision rules). Nevertheless, the continuous-time differential equation framework introduced here remains a powerful analytical lens, especially when combined with modern computational tools.

2.1.8 STRUCTURE OF THE CHAPTER AND NEXT STEPS

Following this introduction, the chapter is divided into several sections:

1. **Review of Ordinary Differential Equations (ODEs):** A succinct overview of first- and higher-order ODEs, linear versus nonlinear forms, and boundary/initial conditions specific to logistics contexts.
2. **Laplace Transform Essentials:** A step-by-step guide to computing Laplace transforms and their inverses, emphasizing the most common functions and signals encountered in logistics (e.g., step demands, impulse orders).
3. **Solving Logistics Problems via Laplace Transforms:** Demonstrations of applying the transform method to solve classic inventory flow and multi-node examples, revealing how to interpret partial fraction decompositions and convolution integrals.
4. **System Stability and Control:** Examination of how poles, zeros, and transfer functions inform the stability of logistics networks under various feedback policies, along with practical guidelines for damping oscillations.
5. **Case Studies in Inventory and Distribution Systems:** Illustrations showing how differential equation models and Laplace methods illuminate real-world challenges, such as a multi-echelon distribution chain facing uncertain demand, or a warehouse implementing a reorder policy with long lead times.

Where relevant, we will connect these concepts to recognized problems in supply chain management (SCM), such as the bullwhip effect or capacity scaling. By grounding the mathematical methods in practical logistics applications, you will not only learn the technical details of differential equations and Laplace transforms but also appreciate their practical usefulness in planning and operating distribution networks.

2.1.9 CONCLUDING REMARKS

Logistics is increasingly recognized as a crucial competitive differentiator for businesses and economies. Whether shipping consumer goods across continents, managing e-commerce fulfillment, or coordinating humanitarian relief efforts, a thorough grasp of the underlying dynamics is essential for success. By embracing mathematical tools like differential equations and Laplace transforms, we gain a deeper and more quantitative perspective on how logistics systems function, adapt, and sometimes fail under stress.

As you progress through this chapter, keep in mind that the power of these methods does not lie in the equations alone, but in the insights they unlock. A well-structured model can help you pinpoint where to place strategic inventory buffers, how to set reorder parameters, or when to scale up transportation capacity. Likewise, a well-chosen control policy, informed by system-theoretic analysis, can tame the bullwhip effect and keep your distribution network stable even in the face of large or unpredictable shifts in demand.

In the years that follow, you may encounter more advanced topics: from nonlinear dynamics and stochastic calculus to machine learning approaches for demand forecasting. Yet, the material here remains foundational. A command of basic differential equations and the use of Laplace transforms paves the way for tackling more complicated problems, whether in continued academic work or in professional practice. We trust that this chapter will provide a solid platform on which to build these further capabilities and an appreciation for the mathematical rigor that underlies efficient, reliable logistics systems.

2.2 REVIEW OF ORDINARY DIFFERENTIAL EQUATIONS (ODES)

Ordinary Differential Equations (ODEs) form the backbone of many models in logistics, enabling us to describe the time evolution of key quantities such as inventory levels, vehicle queues, or capacity utilization. In essence, an ODE expresses a relationship between a function of one independent variable (usually time) and its derivatives. This section presents a concise overview of ODEs, focusing on first- and higher-order equations, their linear versus nonlinear distinctions, and how initial or boundary conditions arise in practical logistics settings.

2.2.1 DEFINITION AND EXAMPLES OF FIRST-ORDER ODES

A first-order ODE involves the function $x(t)$ (which may represent an inventory level, a buffer size, or a throughput rate in logistics) and its first derivative $\frac{dx}{dt}$. The general form of a first-order ODE is

$$\frac{dx}{dt} = f\left(t, x(t)\right),$$

where f is some function of t and $x(t)$. In logistics, one of the simplest examples is the continuous inflow-outflow model. Suppose $I(t)$ denotes the inventory level at time t. We might write

$$\frac{dI}{dt} = \text{Inflow Rate} - \text{Outflow Rate},$$

where Inflow-Rate and Outflow-Rate can be functions of both time and inventory. For instance, if $I(t)$ is replenished at a time-dependent rate $r(t)$ and depleted by a demand $D(t)$, the model could look like

$$\frac{dI}{dt} = r(t) - D(t).$$

This simple equation encapsulates how inventory changes continuously over time as a result of external flows. Solving it might be as straightforward as integrating $r(t) - D(t)$, provided that $I(0)$ (the initial inventory) is known.

2.2.1.1 Initial Conditions in Logistics

To solve a first-order ODE uniquely, we typically need an *initial condition* of the form

$$I(0) = I_0,$$

where I_0 is the known initial inventory at time $t = 0$. In logistics, an initial condition usually corresponds to a starting state: for example, how many units of product a warehouse has at the beginning of the planning horizon. This condition is critical; without I_0, different solutions to the same differential equation might exist, each describing a different initial quantity on hand.

2.2.1.2 Interpretation and Stability

Once an ODE is solved, the solution $I(t)$ provides a continuous-time trajectory of inventory. By examining the solution's behavior, we gain insights into stability. For instance, if $r(t)$ is constant and $D(t)$ is also constant, a solution that tends toward a particular level as $t \to \infty$ indicates a stable steady state (i.e., a point where inflow equals outflow). In contrast, if $r(t)$ or $D(t)$ fluctuates or if the model includes feedback loops (e.g., reorder policies), more complex dynamics may emerge, including oscillations or exponential divergence.

2.2.2 HIGHER-ORDER ODES

While first-order ODEs capture many logistical processes, especially those where the rate of change depends solely on the current state, higher-order equations can be equally pertinent. A *second-order ODE*, for instance, involves the second derivative $\frac{d^2x}{dt^2}$. We might write:

$$\frac{d^2x}{dt^2} = F\left(t, x(t), \frac{dx}{dt}\right).$$

Although second-order equations are most commonly encountered in physical systems (e.g., vibrations or motion dynamics), they can arise in logistics models when there is a delayed or 'inertial' response to changes. For example, consider a warehouse that cannot instantly adjust its reorder rate $Q(t)$. The *adjustment* itself might introduce a second derivative term, capturing the acceleration or deceleration of the production or shipment rate over time.

2.2.2.1 Reduction to First-Order Systems

Any nth-order ODE can be rewritten as a system of n coupled first-order ODEs. This is often done to leverage well-established solution techniques and numerical methods. For example, a second-order ODE

$$\frac{d^2x}{dt^2} = F\left(t, x(t), \frac{dx}{dt}\right)$$

can be decomposed by introducing an auxiliary variable $v(t) = \frac{dx}{dt}$. The system then becomes:

$$\begin{cases} \frac{dx}{dt} = v(t), \\ \frac{dv}{dt} = F\big(t, x(t), v(t)\big). \end{cases}$$

In logistics contexts, such decomposition might be applied when modeling multi-phase flow or complex control strategies where rates of change themselves evolve over time. The decomposition approach is not only a mathematical convenience but also a conceptual one, allowing us to interpret $v(t)$ as an intermediate variable (e.g., the rate of change of inventory or capacity) and $x(t)$ as the fundamental state (e.g., the inventory level).

2.2.2.2 Boundary and Initial Conditions for Higher-Order Problems

Just like first-order ODEs require an initial state, higher-order ODEs demand a corresponding number of initial conditions. For a second-order equation, we commonly specify:

$$x(0) = x_0, \quad \left.\frac{dx}{dt}\right|_{t=0} = v_0.$$

In logistics terms, x_0 might represent the initial inventory, and v_0 the initial rate of change of that inventory (e.g., the ongoing net inflow at $t = 0$). Sometimes, boundary conditions are set up at different times (e.g., $x(0)$ and $x(T)$ for some terminal time T). This can happen when we plan the inventory trajectory over a fixed horizon $[0, T]$ and wish to ensure a certain final state, like leaving minimal leftover stock at $t = T$.

2.2.3 LINEAR VS. NONLINEAR ODES IN LOGISTICS

2.2.3.1 Definition of Linearity

An ODE is called *linear* if the dependent variable and its derivatives appear in a linear fashion; that is, there are no products or nonlinear functions of $x(t)$ or its derivatives. A general nth-order linear ODE can be written as

$$a_n(t)\frac{d^n x}{dt^n} + a_{n-1}(t)\frac{d^{n-1} x}{dt^{n-1}} + \cdots + a_1(t)\frac{dx}{dt} + a_0(t)x(t) = g(t),$$

where $a_0(t), a_1(t), \ldots, a_n(t)$ are known coefficient functions of time, and $g(t)$ is a forcing term.

Logistics often yields *linear* models when the rates of flow or production depend only on time or are proportional to the current state. For instance, a proportional outflow rate $-\alpha x(t)$ is a linear term. Such linear systems are more tractable analytically: solutions can be superimposed (due to linearity), and we can use powerful methods like the Laplace transform to solve them systematically.

2.2.3.2 Nonlinearities in Real Systems

In many practical logistics scenarios, however, *nonlinear* behaviors emerge. Nonlinearity arises if:

1. The rate of change depends on products of variables (e.g., inflow rate depends on both inventory level and an on-hand capacity in a more complex, saturating manner).
2. The system includes step or threshold functions (e.g., reorder policies that suddenly change replenishment rates when inventory dips below a certain level).
3. Constraints induce maximum or minimum flows (e.g., limited transport capacity leading to a piecewise definition of outflow).

These nonlinearities complicate solution approaches. Exact analytical solutions may be elusive, requiring numerical methods or approximations. Yet, they capture real phenomena such as capacity ceilings, lead-time thresholds, and piecewise reorder logic. Thus, while linear ODEs offer a clean theoretical framework and form the basis for classical inventory and queueing analyses, advanced logistics models often require nonlinear formulations to accurately depict real operations.

2.2.4 BOUNDARY AND INITIAL CONDITIONS IN LOGISTICS CONTEXTS

2.2.4.1 Initial Value Problems (IVPs)

Most logistics problems are *Initial Value Problems* (IVPs), where we begin the planning or observation at $t = 0$ with a known system state. For example,

$$\frac{dI}{dt} = r(t) - D(t), \quad I(0) = I_0,$$

is an IVP for a single warehouse. Multi-node networks extend this naturally, where each node's inventory or queue length has an initial condition, reflecting how full each storage location, truck, or container is at the start of the horizon.

2.2.4.2 Boundary Value Problems (BVPs)

Logistics modeling occasionally uses *Boundary Value Problems* (BVPs), particularly in strategic planning. Suppose we want to ensure that our final inventory at time T meets a target I_{target}. One might pose:

$$\frac{d^2x}{dt^2} = F\left(t, x(t), \frac{dx}{dt}\right), \quad x(0) = x_0, \quad x(T) = x_{\text{target}}.$$

If $x(t)$ represents an inventory or supply level across the entire horizon, these boundary conditions specify both the initial level and the final (target) level. This setup is common in scenarios where leftover stock at the end of a season or a project must be carefully controlled.

2.2.5 PRACTICAL CONSIDERATIONS IN ODE MODELING FOR LOGISTICS

2.2.5.1 Model Validation and Parameter Estimation

In any real-world application, the functions and parameters used in an ODE must be estimated from data, expert knowledge, or smaller-scale experiments. For example, a constant outflow rate $D(t)$ could be replaced by a function capturing seasonality: $D(t) = D_0 + D_s \sin(\omega t)$. Calibration involves fitting D_0, D_s, and ω to observed demand patterns. Accurate parameter estimation is vital for producing reliable predictions and for guiding decision-making regarding reorder policies or capacity upgrades.

2.2.5.2 Handling Uncertainties and Stochastic Influences

Though ODEs inherently describe deterministic relationships, real logistics systems face randomness in demand, lead times, and disruptions. Some advanced approaches introduce *stochastic differential equations (SDEs)*, which add noise terms to capture randomness. Others handle uncertainty by treating parameters as random variables or by performing sensitivity analyses to gauge how robust the solution is to parameter changes. Even within a deterministic ODE framework, acknowledging uncertainty can guide decision-makers to implement safety stocks or dynamic reorder thresholds.

2.2.5.3 Numerical Methods and Computability

Finally, even if a logistics ODE is conceptually simple, solutions can be found only through numerical integration in many cases (e.g., *Euler's method*, *Runge–Kutta*, or higher-order schemes). Real-world logistics networks often consist of multiple coupled equations, each representing different echelons in the supply chain. Software implementations (MATLAB, Python, Wolfram Language, etc.) handle these numerical methods effectively, enabling simulation-based approaches for scenario testing and performance evaluations.

2.2.6 CONCLUSION

Ordinary Differential Equations underpin a wide range of logistics models, allowing analysts and practitioners to articulate how system states—like inventories, transportation queues, or production levels—evolve over time. This section has provided a succinct overview of both first- and higher-order ODEs, distinguishing linear from nonlinear forms and highlighting the significance of initial or boundary conditions in real logistics contexts. As we move forward, we will see how these ODEs can be systematically solved using Laplace transforms, yielding further insights into stability, controllability, and responsiveness. Mastering these equations ensures that we can both model and anticipate the dynamics of logistic systems with greater precision, ultimately leading to better-informed decisions in complex supply networks.

2.3 LAPLACE TRANSFORM ESSENTIALS

The Laplace transform is an indispensable tool for analyzing time-domain problems in logistics. By converting ordinary differential equations into an algebraic form in the so-called *s-domain*, the Laplace transform facilitates simpler solutions, especially when dealing with discontinuous or piecewise-defined input signals such as sudden demand spikes (step demands) or instantaneous orders (impulses). This section provides a step-by-step guide to computing Laplace transforms and their inverses, focusing on the functions most frequently encountered in logistics models.

2.3.1 DEFINITION AND BASIC CONCEPTS

The Laplace transform of a function $f(t)$, where $t \geq 0$, is given by the integral

$$F(s) = \mathscr{L}\{f(t)\} = \int_0^\infty e^{-st} f(t)\,dt, \tag{2.1}$$

where s is, in general, a complex variable. In most logistics applications, we consider real values of s large enough to ensure that the integral converges. This convergence typically requires that $f(t)$ not grow faster than e^{at} for some $a \in \mathbb{R}$.

2.3.1.1 Interpretation in Logistics

From a logistics perspective, we can interpret $f(t)$ as any *signal* relevant to system operations: a time-dependent demand rate, $D(t)$; a production or replenishment rate, $Q(t)$; or a scheduling policy that toggles on and off over time. Taking the Laplace transform effectively weighs $f(t)$ by e^{-st}, emphasizing earlier values of $f(t)$ for larger s. This mechanism helps capture *memory* or *accumulated* effects, a crucial aspect in logistics where prior states (e.g., inventory backlogs or lead time queues) can substantially influence future performance.

2.3.2 KEY LAPLACE TRANSFORM PAIRS

A fundamental step in applying the Laplace transform is to recognize certain standard pairs, which can be found in many references or transform tables. Below are several key transform pairs often used to model logistics inputs:

1. **Constant (or Step) Function:**

   $$\mathscr{L}\{1\} = \frac{1}{s} \quad (s > 0).$$

   In many logistics scenarios, a constant demand rate $D(t) = D_0$ is a basic assumption over some interval. The Laplace transform of a constant function is $1/s$, capturing the idea that a continuous, unchanging inflow or outflow dominates behavior at low frequencies (small s).

2. **Unit Step Function (Heaviside Function):**

$$u(t-t_0) = \begin{cases} 0, & t < t_0, \\ 1, & t \geq t_0, \end{cases} \qquad \mathscr{L}\{u(t-t_0)\} = \frac{e^{-st_0}}{s}.$$

This function arises in logistics when a demand or policy *switches on* at a specific time t_0. For example, a step-up in shipping capacity after a system upgrade at t_0 can be modeled as $Q(t) = Q_0\, u(t-t_0)$.

3. **Exponentials:**

$$\mathscr{L}\{e^{at}\} = \frac{1}{s-a} \quad (s > a).$$

Exponential forms appear frequently when modeling decay toward equilibrium (e.g., depletion of inventory under a simple proportional outflow) or growth (e.g., backlog accumulation under certain feedback loops).

4. **Dirac Delta (Impulse):**

$$\delta(t-t_0), \quad \mathscr{L}\{\delta(t-t_0)\} = e^{-st_0}.$$

In logistics, $\delta(t-t_0)$ can represent a sudden, *instantaneous* order or supply shock at time t_0. For instance, a bulk one-time demand for spare parts might be approximated by a Dirac delta if it occurs within a very narrow window compared to the overall planning horizon.

5. **Polynomials:**

$$\mathscr{L}\{t^n\} = \frac{n!}{s^{n+1}},$$

which may appear in stylized models of time-varying demand (e.g., a demand ramp that grows linearly or quadratically with time).

Recognizing these pairs is essential, as real-world signals in logistics are often constructed from piecewise combinations of step, impulse, or exponential terms.

2.3.2.1 Superposition and Linearity

One of the great advantages of the Laplace transform is its linearity:

$$\mathscr{L}\{af(t) + bg(t)\} = aF(s) + bG(s),$$

where a and b are constants, $f(t)$ and $g(t)$ are functions, and $F(s)$ and $G(s)$ are their respective transforms. If demand $D(t)$ in a logistics system is the sum of a baseline constant demand D_0 and an additional step demand that commences at t_1, we can write

$$D(t) = D_0 + D_{\text{step}}\, u(t-t_1),$$

and easily take the Laplace transform by adding the transforms of D_0 and $D_{\text{step}}\, u(t-t_1)$.

2.3.3 PROPERTIES OF THE LAPLACE TRANSFORM IN LOGISTICS CONTEXTS

2.3.3.1 Transform of a Derivative

One of the most powerful properties for solving differential equations is:

$$\mathscr{L}\left\{\frac{df(t)}{dt}\right\} = sF(s) - f(0),$$

where $F(s) = \mathscr{L}\{f(t)\}$. More generally, for the nth derivative:

$$\mathscr{L}\left\{\frac{d^n f(t)}{dt^n}\right\} = s^n F(s) - s^{n-1} f(0) - s^{n-2} \frac{df(t)}{dt}\bigg|_{t=0} - \cdots - \frac{d^{n-1} f(t)}{dt^{n-1}}\bigg|_{t=0}.$$

In the logistics setting, if $I(t)$ is inventory, then $\frac{dI}{dt}$ might be the net flow (inflow minus outflow). Taking the Laplace transform of $\frac{dI}{dt}$ turns differentiation into an algebraic factor s, drastically simplifying solution procedures for linear ODEs.

2.3.3.2 Convolution Theorem

Convolution naturally arises in logistics when modeling how one node's output becomes another node's input with a certain transport or lead time. The convolution theorem states:

$$\mathscr{L}\{f(t) * g(t)\} = F(s)G(s),$$

where the convolution $(f * g)(t)$ is defined by

$$(f * g)(t) = \int_0^t f(\tau)g(t-\tau)\,d\tau.$$

If $f(t)$ represents an "input signal" (such as raw materials shipped out of a supplier) and $g(t)$ describes a "transfer function" (such as a delay or a capacity filter), then $f(t) * g(t)$ tells us how the output to the next stage arrives or accumulates over time. In the s-domain, the multiplication $F(s)G(s)$ is typically easier to manage than performing convolution integrals directly in the time domain.

2.3.3.3 Initial and Final Value Theorems

Two additional results that are quite helpful in logistics analysis, especially for steady-state or near-initial responses, are:

1. **Initial Value Theorem:**

$$\lim_{t \to 0^+} f(t) = \lim_{s \to \infty} sF(s),$$

which provides $f(0^+)$ directly from $F(s)$ if the limit exists. For instance, we can use this to confirm an initial inventory level when taking a transform-based approach.

2. **Final Value Theorem:**

$$\lim_{t\to\infty} f(t) = \lim_{s\to 0^+} sF(s),$$

useful for finding the long-term behavior of an inventory level, backlog, or any other logistic variable of interest. This theorem helps to check whether a system settles at a stable steady-state or diverges.

2.3.4 INVERSE LAPLACE TRANSFORMS AND PARTIAL FRACTION DECOMPOSITION

Once an ODE is mapped to an algebraic equation via the Laplace transform, we typically solve for the unknown $F(s)$ in the s-domain. However, to interpret the solution in physical terms, we must retrieve $f(t)$ by applying the *inverse* Laplace transform:

$$f(t) = \mathscr{L}^{-1}\{F(s)\}.$$

For many logistic models, especially linear ones, the solution in the s-domain takes the form of a rational function:

$$F(s) = \frac{P(s)}{Q(s)},$$

where $P(s)$ and $Q(s)$ are polynomials in s. One standard technique to invert such expressions is *partial fraction decomposition*, which rewrites $F(s)$ as a sum of simpler terms whose inverse transforms are known from tables or standard formulas.

For example, if we have:

$$F(s) = \frac{3}{(s+1)(s+2)},$$

we decompose it as

$$F(s) = \frac{A}{s+1} + \frac{B}{s+2}.$$

Multiplying both sides by $(s+1)(s+2)$ gives $3 = A(s+2) + B(s+1)$. We determine A and B by choosing convenient values of s:

- Set $s = -1$: $3 = A(-1+2) + B(0) \implies 3 = A(1) \implies A = 3.$
- Set $s = -2$: $3 = A(0) + B(-2+1) \implies 3 = B(-1) \implies B = -3.$

Hence the decomposition is

$$F(s) = \frac{3}{s+1} - \frac{3}{s+2}.$$

Taking inverse Laplace transforms term by term,

$$f(t) = \mathscr{L}^{-1}\{F(s)\} = 3e^{-1\cdot t} - 3e^{-2\cdot t} = 3\left(e^{-t} - e^{-2t}\right).$$

In a logistics scenario, these exponentials can represent decaying or growing contributions to a variable such as inventory, due to the interplay of inflows and outflows governed by the underlying ODE.

2.3.5 ILLUSTRATIVE EXAMPLES OF LOGISTICS SIGNALS

2.3.5.1 Step Demand

A sudden, permanent jump in demand at $t = 0$ can be represented by

$$D(t) = D_0\, u(t),$$

where $u(t)$ is the Heaviside step function (1 for $t \geq 0$, 0 for $t < 0$). Its Laplace transform is

$$\mathscr{L}\{D_0\, u(t)\} = \frac{D_0}{s}.$$

This forms the basis for analyzing how a warehouse or distribution network responds to an instantaneous but sustained increase in demand. The outcome might be an exponential adjustment to a new inventory level, or it might drive the system into an unstable oscillation if feedback policies are not well-tuned.

2.3.5.2 Impulse Orders

An unexpected one-time bulk order at $t = t_0$ can be modeled by

$$D(t) = D_{\text{imp}}\, \delta(t - t_0),$$

where $\delta(\cdot)$ is the Dirac delta. The Laplace transform is

$$\mathscr{L}\{D_{\text{imp}}\, \delta(t - t_0)\} = D_{\text{imp}}\, e^{-st_0}.$$

Solving an ODE subject to this impulse allows logisticians to see how the system will *pulse* in response (e.g., a temporary shortage or sudden queue buildup) before returning to nominal conditions. For time-critical operations—such as just-in-time manufacturing or perishable goods handling—understanding this response is essential to design adequate buffer stocks.

2.3.6 PRACTICAL CONSIDERATIONS AND SUMMARY

2.3.6.1 Why Use the Laplace Transform for Logistics?

In continuous-time logistics models, we often deal with nontrivial initial conditions, diverse input signals (step, ramp, pulse), and coupled differential equations representing multi-echelon supply chains. The Laplace transform streamlines the solution process by:

1. Converting derivatives into algebraic terms (s^n factors).
2. Providing a straightforward mechanism (partial fraction decomposition) to invert solutions into the time domain.
3. Letting us exploit well-known theorems (convolution, initial/final values) to interpret system performance quickly.

2.3.6.2 Numerical Tools

While many Laplace transforms and inverse transforms can be handled analytically, complex or piecewise inputs may require symbolic or numerical tools. Software packages (e.g., MATLAB, Python with `Sympy`, Wolfram Language) can perform transforms automatically, facilitating rapid prototyping and scenario testing. Such tools are especially helpful if the logistic system is large-scale, comprising multiple stages with lead times, capacity constraints, and distinct operational modes.

2.3.6.3 Conclusion and Next Steps

Understanding the Laplace transform is vital for anyone seeking to derive and analyze ODE-based models in logistics, as it simplifies the solution and clarifies how step demands, impulses, and other signals drive system dynamics. Equipped with the transform properties, standard pairs, and inverse transforms, we can now tackle the typical equations encountered in inventory management, distribution flows, and beyond. In the upcoming sections, we will delve into applying these methods to actual logistics problems, demonstrating how to solve inventory flow ODEs and interpret the resulting time-domain behaviors in practical, decision-making contexts.

2.4 SOLVING LOGISTICS PROBLEMS VIA LAPLACE TRANSFORMS

Having introduced both Ordinary Differential Equations (ODEs) and the fundamentals of Laplace transforms, we now turn our attention to applying these tools in concrete logistics contexts. In this section, we will demonstrate how the transform method streamlines the solution of classic inventory flow models and multi-node distribution systems. We will explore how partial fraction decomposition and convolution integrals arise naturally, and we will interpret the results in the time domain to draw meaningful insights into logistics performance.

2.4.1 A CLASSIC SINGLE-NODE INVENTORY MODEL

Let us begin with a simple yet illustrative scenario: a warehouse or distribution center whose inventory level $I(t)$ is governed by the linear ODE

$$\frac{dI}{dt} + \alpha I(t) = D(t), \tag{2.2}$$

where $\alpha > 0$ is a constant rate parameter capturing, for example, the proportional outflow relative to on-hand inventory (common in continuous-review systems), and $D(t)$ is an external demand or replenishment signal. Assume we have the initial condition

$$I(0) = I_0.$$

Our goal is to determine $I(t)$ for all $t \geq 0$.

2.4.1.1 Taking the Laplace Transform

From Section 2.3, we recall that taking the Laplace transform of both sides of (2.2) transforms differentiation into an algebraic term. Specifically, we apply $\mathscr{L}\{\cdot\}$ to each term:

$$\mathscr{L}\left\{\frac{dI}{dt}\right\} + \alpha\mathscr{L}\{I(t)\} = \mathscr{L}\{D(t)\}.$$

Using the derivative property $\mathscr{L}\{\frac{dI}{dt}\} = sI(s) - I(0)$, we obtain

$$\left(sI(s) - I_0\right) + \alpha I(s) = D(s),$$

where $I(s) = \mathscr{L}\{I(t)\}$ and $D(s) = \mathscr{L}\{D(t)\}$.

Rearranging for $I(s)$, we find

$$I(s) = \frac{I_0 + D(s)}{s+\alpha}. \tag{*}$$

2.4.1.2 Interpreting and Inverting the Laplace-Domain Solution

Equation $(*)$ shows that $I(s)$ is the product of two distinct terms:

$$I(s) = \left(\frac{I_0}{s+\alpha}\right) + \left(\frac{D(s)}{s+\alpha}\right).$$

Hence, we can split the solution into a *homogeneous response* (driven by initial inventory I_0) and a *particular response* (driven by the input $D(t)$):

1. **Homogeneous Response (Initial Inventory):** $\mathscr{L}^{-1}\left\{\frac{I_0}{s+\alpha}\right\} = I_0\, e^{-\alpha t}$, which implies that, in the absence of further inflows or outflows, the inventory (or system state) would decay exponentially at rate α.
2. **Particular Response (Demand/Replenishment Input):** $\mathscr{L}^{-1}\left\{\frac{D(s)}{s+\alpha}\right\} = (d * e^{-\alpha t})(t)$, where $d(t) = \mathscr{L}^{-1}\{D(s)\} = D(t)$. The notation $*$ indicates *convolution*, which arises by the Convolution Theorem:

$$\mathscr{L}^{-1}\left\{\frac{D(s)}{s+\alpha}\right\} = \mathscr{L}^{-1}\{D(s)\} * \mathscr{L}^{-1}\{\tfrac{1}{s+\alpha}\}.$$

Since $\mathscr{L}^{-1}\{\frac{1}{s+\alpha}\} = e^{-\alpha t}\, u(t)$, we get:

$$(D * e^{-\alpha\cdot})(t) = \int_0^t D(\tau)\, e^{-\alpha(t-\tau)}\, d\tau.$$

In logistics terms, this means every past increment of demand $D(\tau)$ exerts an effect that *decays exponentially* over the time interval (τ, t), reflecting how system responses to earlier demands gradually diminish if α is associated with outflow or service.

Putting these pieces together,

$$I(t) = I_0 e^{-\alpha t} + \int_0^t D(\tau)\, e^{-\alpha(t-\tau)}\, d\tau.$$

This expression is the *inverse Laplace transform* of $I(s)$. It succinctly reveals how the system's inventory is a blend of the memory of its initial state and the cumulative, exponentially weighted effect of all past demand events.

2.4.1.3 A Step Demand Example

As a classic example, let us set $D(t) = D_0\, u(t)$ (a constant demand starting at $t = 0$). Its Laplace transform is $D(s) = D_0/s$. Substituting into $(*)$ gives

$$I(s) = \frac{I_0}{s+\alpha} + \frac{D_0/s}{s+\alpha}.$$

A partial fraction approach can be used on the second term:

$$\frac{D_0/s}{s+\alpha} = \frac{D_0}{s(s+\alpha)}.$$

Decomposing,

$$\frac{D_0}{s(s+\alpha)} = \frac{A}{s} + \frac{B}{s+\alpha},$$

we can solve for A and B. A quick computation yields $A = \frac{D_0}{\alpha}$ and $B = -\frac{D_0}{\alpha}$. Hence,

$$\frac{D_0}{s(s+\alpha)} = \frac{D_0/\alpha}{s} - \frac{D_0/\alpha}{s+\alpha}.$$

The inverse Laplace transforms are straightforward:

$$\mathcal{L}^{-1}\left\{\frac{1}{s}\right\} = 1, \quad \mathcal{L}^{-1}\left\{\frac{1}{s+\alpha}\right\} = e^{-\alpha t}.$$

Consequently,

$$I(t) = I_0 e^{-\alpha t} + \frac{D_0}{\alpha}\left(1 - e^{-\alpha t}\right).$$

As $t \to \infty$, the system converges to $I(\infty) = \frac{D_0}{\alpha}$. Interpreting this in logistics terms: in the long run, if the net outflow is $\alpha I(t)$ and the constant inflow (or demand offset) is D_0, the stable equilibrium of inventory is where $D_0 = \alpha I(\infty)$. The solution thus directly matches our intuition about how a simple linear system balances supply and demand over time.

2.4.2 A TWO-NODE FLOW EXAMPLE

Many logistics networks involve more than one location or node. Suppose we have a simple two-node system:

$$\begin{cases} \frac{dI_1}{dt} = -\beta I_1(t) + F(t), \\ \frac{dI_2}{dt} = \beta I_1(t) - \alpha I_2(t), \end{cases}$$

where:

1. $I_1(t)$ is the inventory (or queue size) at node 1,
2. $I_2(t)$ is the inventory (or queue size) at node 2,
3. $\beta > 0$ is the flow rate from node 1 to node 2 (proportional to $I_1(t)$ if node 1 is feeding node 2),
4. $\alpha > 0$ is the outflow rate from node 2, e.g., final demand consumption,
5. $F(t)$ is an external replenishment or demand offset arriving at node 1.

Assume initial conditions $I_1(0) = I_{1,0}$ and $I_2(0) = I_{2,0}$.

2.4.2.1 Transforming Each ODE

Take the Laplace transform of each equation. Let

$$I_1(s) = \mathscr{L}\{I_1(t)\}, \quad I_2(s) = \mathscr{L}\{I_2(t)\}, \quad F(s) = \mathscr{L}\{F(t)\}.$$

For the first equation:

$$sI_1(s) - I_{1,0} = -\beta I_1(s) + F(s).$$

Rearranging,

$$(s+\beta) I_1(s) = I_{1,0} + F(s).$$

Thus,

$$I_1(s) = \frac{I_{1,0} + F(s)}{s+\beta}.$$

For the second equation:

$$sI_2(s) - I_{2,0} = \beta I_1(s) - \alpha I_2(s).$$

Hence,

$$sI_2(s) + \alpha I_2(s) = I_{2,0} + \beta I_1(s).$$

Or,

$$I_2(s) = \frac{I_{2,0} + \beta I_1(s)}{s+\alpha}.$$

2.4.2.2 Coupling and Substitution

To eliminate $I_1(s)$ in (2.4.2.1), we substitute the expression from (2.4.2.1):

$$I_2(s) = \frac{I_{2,0} + \beta \dfrac{I_{1,0} + F(s)}{s+\beta}}{s+\alpha} = \frac{I_{2,0}(s+\beta) + \beta\left(I_{1,0} + F(s)\right)}{(s+\alpha)(s+\beta)}.$$

Thus, we have explicit expressions for both $I_1(s)$ and $I_2(s)$ in terms of $F(s)$ and the known constants/initial conditions:

$$I_1(s) = \frac{I_{1,0} + F(s)}{s+\beta}, \qquad I_2(s) = \frac{I_{2,0}(s+\beta) + \beta I_{1,0} + \beta F(s)}{(s+\alpha)(s+\beta)}.$$

2.4.2.3 Partial Fractions and Convolution Interpretation

We can further simplify these expressions using partial fraction decomposition, especially for the term involving $F(s)/(s+\beta)$ and $\left((s+\beta)/(s+\alpha)(s+\beta)\right)$. The resulting time-domain solutions will be sums and convolutions of exponentials multiplied by the inverse Laplace transforms of $F(s)$. In logistic terms:

1. $I_1(t)$ is the superposition of how initial inventory $I_{1,0}$ decays at rate β, plus a convolution of the external signal $F(t)$ with $e^{-\beta t}$.
2. $I_2(t)$ similarly depends on both its own initial inventory $I_{2,0}$ and the stream $\beta I_1(t)$ that feeds it, which then decays at rate α.

The two-node example highlights how the Laplace transform handles coupled equations: by turning them into algebraic forms that reveal the network's structure more transparently. Once we invert to the time domain, we can analyze how changes in $F(t)$ (such as step or impulsive arrivals) percolate through node 1 to node 2, possibly generating oscillatory or exponential-like transitions until a new equilibrium is reached.

2.4.3 PRACTICAL INSIGHTS AND INTERPRETATION

2.4.3.1 System Response and Transient Behavior

The step-by-step Laplace transform solution uncovers key properties of logistics systems:

1. **Transient Responses:** Exponentially decaying terms (e.g., $e^{-\alpha t}$ or $e^{-\beta t}$) dominate the early transient phase. By examining poles ($s = -\alpha$, $s = -\beta$), we gauge how quickly each node sheds or responds to perturbations.
2. **Steady States:** In linear systems with constant inputs, the final value ($\lim_{t\to\infty} I(t)$) emerges clearly from the partial fraction decomposition and the Final Value Theorem. Understanding these equilibria is crucial for setting capacity or safety stock levels.

3. **Feedback Loops:** In more complex models, reorder policies or capacity saturation constraints can introduce nonlinearities or delays. While these are often tackled with modifications to the transforms or with piecewise definitions, the principle remains: transform the ODEs, solve algebraically, invert back, and interpret.

2.4.3.2 Advantages Over Direct Time-Domain Solutions

In many logistical ODEs, direct time-domain solution can be cumbersome, particularly if $D(t)$ or $F(t)$ includes step changes, impulses, or piecewise functions. The Laplace approach consolidates each piece of the problem (initial conditions, dynamic terms, boundary conditions) into an algebraic form. The partial fraction decomposition then simplifies the inverse transform to standard exponentials and polynomials, a process often more straightforward than integrating piecewise ODEs separately.

2.4.3.3 Extensions to Larger Networks and Lead Times

Though we showcased a two-node system, realistic logistics networks involve multiple echelons or parallel paths. The same procedure generalizes: each node has an ODE representing its net flow, leading to a set of coupled linear (or linearized) equations in the s-domain. Additionally, when lead times or shipping delays are significant, one can incorporate time-delay operators, which in the Laplace domain appear as factors $e^{-s\tau}$. Convolution integrals then interpret how past shipments "arrive" at downstream nodes after delay τ.

2.4.4 CONCLUDING REMARKS ON THE LAPLACE TRANSFORM METHOD

The examples in this section show how systematically applying Laplace transforms to logistics ODEs offers a clear path from differential equations to explicit, interpretable solutions. We demonstrated:

1. **Single-Node Inventory Flows:** How partial fraction decomposition neatly handles step demands or impulses, clarifying the equilibrium and transient behavior.
2. **Multi-Node Coupling:** How to solve a system of ODEs algebraically, then invert to the time domain to understand flow propagation in a mini-network.
3. **Convolution Insights:** How each node's response can be seen as the convolution of incoming signals with decaying exponential kernels, highlighting how past inflows gradually lose impact over time.

With these foundations, you are now prepared to tackle even more complex scenarios, such as networks with multiple distribution centers, capacity constraints, or feedback reorder policies. In the next sections, we will look at *system stability* and *control strategies*, focusing on how to keep inventory or queue levels within acceptable bounds and how to design robust policies to handle uncertain or fluctuating demands.

2.5 SYSTEM STABILITY AND CONTROL

Stability is a central concern when modeling logistics systems. Even well-intentioned adjustments in production rates or reorder policies can lead to pronounced oscillations in inventory and delivery flows, ultimately increasing costs or degrading service levels. By analyzing the *poles*, *zeros*, and *transfer functions* of linearized models, logisticians and engineers can ascertain whether a system will converge to a steady state, exhibit oscillatory behavior, or diverge entirely. This section surveys these foundational concepts, demonstrates their application in feedback-based logistics, and offers practical guidelines for damping unwanted fluctuations.

2.5.1 TRANSFER FUNCTIONS IN LOGISTICS NETWORKS

A transfer function characterizes how an output signal responds to an input signal in the Laplace domain. For a single-node inventory example, we might define the transfer function from demand $D(t)$ to inventory $I(t)$ by

$$G(s) = \frac{I(s)}{D(s)},$$

provided the system is linear and time-invariant. Typically, $G(s)$ captures how the node's inventory evolves in response to changes in demand.

In multi-node logistics networks, one can derive analogous transfer functions to describe how inputs (such as external demand, reorder signals, or production instructions) filter through multiple warehouses or distribution centers, ultimately affecting inventory levels, lead times, or backlogs.

2.5.2 POLES, ZEROS, AND THEIR ROLE IN STABILITY

Poles of a transfer function $G(s)$ are the values of s for which $Q(s) = 0$. Zeros are values of s for which $P(s) = 0$. From a dynamical perspective:

1. **Poles** generally dictate the system's natural modes or characteristic responses. In continuous-time models, if any pole has a positive real part, the system's response can grow unbounded, indicating *instability*. Poles on the imaginary axis (i.e., $\text{Re}(s) = 0$) typically introduce sustained oscillations. Poles in the left half-plane (i.e., $\text{Re}(s) < 0$) indicate exponential decay of those modes, which fosters stability.
2. **Zeros** can alter how input signals pass through the system. While they do not alone determine stability, zeros can change the amplitude or phase of responses in specific frequency ranges. In some feedback configurations, zeros can partially cancel poles or contribute to phenomena like overshoot or undershoot.

In logistics contexts, poles often arise from parameters like inventory outflow rate (α), inter-node flow rates (β_i), or reorder feedback gains. Large positive gains can push poles rightward into the right half-plane, leading to oscillations or even runaway divergence. By contrast, well-chosen gains help shift poles leftward, yielding quicker yet stable convergence to a target inventory level.

2.5.3 FEEDBACK POLICIES AND THE BULLWHIP EFFECT

A hallmark of many supply chains is the *bullwhip effect*: small variations in consumer demand cause disproportionately large oscillations in upstream production and inventory. One major driver is the tendency for each echelon to overreact to changes, often via feedback loops.

2.5.3.1 Example: Proportional Reorder Policy

Consider a single-node system in which inventory $I(t)$ is continuously compared to a target level I_{target}. The node adjusts its incoming replenishment rate $R(t)$ proportionally to the discrepancy:

$$R(t) = k\left(I_{\text{target}} - I(t)\right),$$

where $k > 0$ is the proportional feedback gain. In the simplest linear model, net flow into the node might be

$$\frac{dI}{dt} = R(t) - D(t) = k\left(I_{\text{target}} - I(t)\right) - D(t).$$

Rearranging,

$$\frac{dI}{dt} + kI(t) = kI_{\text{target}} - D(t).$$

Taking the Laplace transform and solving for $I(s)$:

$$sI(s) - I(0) + kI(s) = kI_{\text{target}}\frac{1}{s} - D(s).$$

Hence,

$$I(s) = \frac{I(0) + kI_{\text{target}}\frac{1}{s} - D(s)}{s+k}.$$

From this expression, we see a single pole at $s = -k$ (plus potential poles from $D(s)$). Because $k > 0$, $s = -k$ lies in the left half-plane, suggesting the system is stable in the sense of asymptotic decay (no free oscillations). However, real supply chains frequently exhibit time delays (transport, lead times) or multi-echelon feedback loops that can introduce more intricate pole locations. If the effective feedback loop is too aggressive, the real parts of some poles may shift rightward, leading to oscillations or instability.

2.5.4 GUIDELINES FOR DAMPING OSCILLATIONS IN LOGISTICS SYSTEMS

2.5.4.1 Pole Placement and Controller Tuning

One of the most direct ways to ensure stability is to *place the poles* of the logistics system in the left half-plane by tuning feedback gains. A low gain k might yield minimal oscillations but slow response (meaning inventory can remain off-target for extended periods). A high k improves responsiveness but risks overshoot and oscillations, especially when lead times enter the picture. A middle ground that yields *critical damping* (no oscillation, fastest convergence) or *underdamping* (some controlled oscillation) may be most effective.

In multi-echelon or multi-node environments, each node's reorder policy can be seen as a local controller. Setting consistent, coordinated gains across all nodes helps avoid destructive interaction, where one node's surge triggers another node's overreaction.

2.5.4.2 Time Delays and the Root Locus

With nonzero lead times τ, the system often includes factors like $e^{-s\tau}$ in the transfer function. These delay terms can introduce infinitely many poles as expansions in the s-plane. While exact analysis may become challenging, *root locus* or *Nyquist* methods can still guide design choices. In essence, for each potential value of k, one traces how poles move in the complex plane. If any path crosses the imaginary axis, instability arises. Mitigating strategies include splitting replenishment into smaller, more frequent orders (reducing effective delays), or implementing forecast smoothing policies.

2.5.4.3 Filtering and Inventory Smoothing

Another broad approach to mitigating the bullwhip effect and damping oscillations is *demand filtering*. Instead of reacting immediately and fully to each demand fluctuation, managers may deploy smoothing techniques such as moving averages or exponential smoothing. From a Laplace-transform perspective, this smoothing is akin to adding low-pass filters (factors like $\frac{1}{\alpha s+1}$) to the feedback loop. These filters dampen high-frequency components of demand signals, preventing abrupt spikes in orders upstream.

While this can improve stability, it also slows responsiveness. As with gains k, finding an optimal degree of smoothing involves balancing inventory cost against the risk of short-term stockouts when demand spikes sharply.

2.5.5 ZEROS AND OVERSHOOT IN LOGISTICS

In certain logistic systems, zeros in the transfer function can create *overshoot* or *undershoot*. For instance, if a node has an upstream capacity limit that triggers a piecewise or saturating behavior, the linearized model may exhibit zeros that alter transient dynamics. Although zeros do not dictate exponential growth or decay on their own, they impact how the output amplitude evolves in response to a step or impulse. In practice, if you notice that your inventory overshoots its target level, rechecking the system's zeros (and potential zero-pole cancellations) might illuminate an overly sensitive or misguided control logic.

2.5.6 CASE ILLUSTRATION: TWO-ECHELON NETWORK WITH FEEDBACK

Imagine a manufacturer ($node_1$) feeding a warehouse ($node_2$) that meets external demand $D(t)$. Both nodes adjust their production or reorder rates via proportional control based on a target level. Writing down the differential equations for $I_1(t)$ and $I_2(t)$, we see that each node's outflow depends not only on its own inventory deviation but also on the upstream or downstream signals. In the s-domain, we obtain a set of coupled transfer functions $G_{11}(s), G_{12}(s), G_{21}(s), G_{22}(s)$ linking demands, reorders, and inventories. Analyzing the resultant characteristic equation $Q(s) = 0$, we locate poles. If these poles are complex with large positive real parts, we expect pronounced oscillations or potential divergence. Adjusting each node's gain or adding a small filter at either node can shift poles left, thereby moderating swings.

2.5.7 PRACTICAL STABILITY GUIDELINES

To summarize:

1. **Assess Pole Locations:** After setting up the linearized model, check the real parts of the poles. Ensure that all are negative for stable operation.
2. **Tune Gains Moderately:** Increasing feedback gain improves responsiveness but risks overshoot and instability if lead times or multi-echelons exist. Use moderate gains or phased adjustments rather than abrupt, large changes.
3. **Incorporate Smoothing:** Employ demand smoothing or forecast filtering to quell high-frequency noise. This reduces the amplitude of oscillatory modes but may increase the lag in responding to real shifts.
4. **Coordinate Across Nodes:** In multi-node systems, align control policies so that one node's push or pull signals do not amplify upstream or downstream instabilities.
5. **Analyze Delays Carefully:** Lead times transform stable poles into near-borderline or unstable poles. Root locus or frequency-domain analyses help predict when a small gain change can create big swings.
6. **Simulate Nonlinear Effects:** Real logistics processes (capacity constraints, threshold reorder triggers) introduce nonlinearities. Even if you design gains via a linear approximation, use simulation to confirm stable operation under actual constraints.

2.5.8 CONCLUSION AND OUTLOOK

Poles, zeros, and transfer functions provide a powerful lens for understanding the dynamic behavior of logistics networks under feedback control. By examining the location of poles in the complex plane, one can judge whether inventory or production systems settle swiftly to targets, wobble, or spiral out of control. Zeros further shape the form of transient responses, influencing whether changes in demand cause overshoot, undershoot, or near-perfect tracking.

In practice, controlling a logistics network for stability is as much an art as a science: theoretical pole placement must square with real-world constraints and uncertainties. Nevertheless, adhering to the guidelines discussed here—moderating feedback gains, smoothing abrupt demand fluctuations, and coordinating multi-echelon policies—goes a long way toward avoiding costly bullwhip effects and achieving robust, efficient supply chain operations.

Moving forward, we can extend these stability insights to broader problem domains, including *robust control* (handling parameter uncertainties) and *optimal control* (balancing multiple performance criteria). These advanced topics build on the same transfer function framework introduced here but incorporate additional layers of complexity to handle the full range of challenges in modern logistics.

2.6 SYNTHESIZING CONCEPTS FOR LARGER NETWORKS

Thus far, our focus has been on models with one or two nodes, illustrating fundamental techniques in differential equations and Laplace transforms to capture the flow of goods and information. Real-world logistics, however, often spans multiple echelons and a vast network of distribution centers, manufacturing plants, and transportation links. This section explores how the mathematical foundations we have developed scale up to represent and analyze such complex systems. We will highlight the system-level implications of coupling many nodes and discuss the challenges and opportunities in applying these methods to genuine supply chain operations. In a later chapter on *Network Science in Logistics*, we will examine more specialized analyses of network topologies and their role in connectivity, robustness, and sustainability.

2.6.1 FROM FEW NODES TO MANY: THE GROWTH OF COMPLEXITY

When moving from single-node or two-node models to larger, multi-echelon networks, several aspects of the problem grow in intricacy:

1. **Dimension of State Space:** In a network of N facilities, each with its own inventory level $I_i(t)$, we can end up with a coupled system of N differential equations:
$$\frac{d\mathbf{I}(t)}{dt} = \mathbf{f}(\mathbf{I}(t), t),$$
where $\mathbf{I}(t) = (I_1(t), I_2(t), \ldots, I_N(t))^\top$. This state vector may also include transportation pipelines (in-transit stock) or queue lengths at distribution centers. The dimensionality of the system can become large, and even apparently small feedback loops can profoundly influence dynamics.
2. **Complex Flows and Constraints:** Real supply chains often feature multi-directional flows. Raw materials flow from upstream suppliers to factories, finished goods might flow through regional distribution centers, returns or reverse flows might head back for remanufacturing or recycling, and lateral transshipments can occur between facilities. Representing these flows mathematically requires careful bookkeeping of flow rates and constraints, leading to more sophisticated ODEs or sets of difference equations for each arc.

3. **Diverse Demand Signals:** Each facility might face distinct demand patterns—some more predictable (steady orders from a small set of customers) and others highly variable (online retail, seasonal peaks). The overall network is then exposed to a constellation of inputs $D_1(t), D_2(t), \dots, D_N(t)$, each with its own forecast quality and uncertainty. Coupling emerges if one node's reorder policy triggers a surge of replenishment requests that cascade upstream or compete for limited transportation capacity.

2.6.2 CONSTRUCTING AND SOLVING NETWORK-WIDE MODELS

2.6.2.1 Matrix Formulations and the Laplace Domain

One can extend the Laplace transform approach to multi-node networks by moving to a *state-space* or *matrix* representation. Suppose we have a linearized network:

$$\frac{d\mathbf{I}(t)}{dt} = \mathbf{A}\mathbf{I}(t) + \mathbf{B}\mathbf{D}(t),$$

where $\mathbf{I}(t)$ is an $N \times 1$ state vector of inventories, $\mathbf{D}(t)$ is an $M \times 1$ vector of exogenous demands (or other inputs), $\mathbf{A}$ is an $N \times N$ system matrix describing the flow or feedback between nodes, and $\mathbf{B}$ is an $N \times M$ input matrix. Taking the Laplace transform of both sides,

$$s\mathbf{I}(s) - \mathbf{I}(0) = \mathbf{A}\mathbf{I}(s) + \mathbf{B}\mathbf{D}(s).$$

Rearranging,

$$\left(s\mathbf{I} - \mathbf{A}\right)\mathbf{I}(s) = \mathbf{I}(0) + \mathbf{B}\mathbf{D}(s).$$

Then,

$$\mathbf{I}(s) = \left(s\mathbf{I} - \mathbf{A}\right)^{-1}\left[\mathbf{I}(0) + \mathbf{B}\mathbf{D}(s)\right].$$

The inverse Laplace transform of this expression yields the time-domain solution. However, the inverse operation may be much more involved than in the single-node case, since $(s\mathbf{I} - \mathbf{A})^{-1}$ is typically a rational matrix whose determinant can have a high degree. Nonetheless, modern symbolic or numerical tools can handle many problems of practical size.

2.6.2.2 Handling Coupled Constraints and Nonlinearities

In real logistics networks, capacity constraints, reorder thresholds, or saturations can introduce nonlinearities that complicate the direct use of Laplace transforms. One approach is to *linearize* around an operating point (e.g., a steady flow or equilibrium inventory vector) to analyze local stability and small-signal response. This yields a linear time-invariant (LTI) approximation valid near the operating state. For large deviations, piecewise approximations or fully numerical methods (e.g., time-stepping simulations) might be employed.

2.6.3 SYSTEM-LEVEL DYNAMICS AND EMERGING PHENOMENA

A fundamental insight from networked systems is that local feedback loops can combine in unexpected ways, giving rise to global phenomena. Two commonly encountered dynamics in multi-node supply chains are:

1. **Oscillatory Amplification (Extended Bullwhip):** If each node in a chain implements a high-gain reorder policy without coordinating with upstream or downstream nodes, fluctuations in demand at one end can reverberate with increasing amplitude as they travel across the network. Poles corresponding to the collective system can shift dangerously close to the imaginary axis, causing persistent swings in multiple echelons.
2. **Bottleneck Effects and Time-Varying Flows:** In a more general topology (e.g., a distribution hub that feeds multiple regional centers), local capacity constraints can ripple out. When one node saturates, it effectively clips the flow to downstream nodes. This can lead to nonlinear wave-like effects, where backlogs build behind the constrained node and starve or starve-then-flood other nodes in cyclical patterns.

Analyzing these phenomena through the lens of ODEs and Laplace transforms provides a structured way to identify which parameters (flow coefficients, lead times, reorder gains) play the largest role in dampening or amplifying instabilities.

2.6.4 IMPLICATIONS FOR REAL-WORLD SUPPLY CHAINS

2.6.4.1 Scalability and Computation

For networks with tens, hundreds, or even thousands of nodes—each with its own local control logic—direct analytical solutions become cumbersome. In many practical cases, planners resort to:

1. **High-Level Aggregation:** Grouping certain nodes into macro-nodes or "virtual facilities" to reduce dimensionality. Aggregation can preserve key dynamics if done carefully but may omit important details such as local constraints or regional lead times.
2. **Numerical Simulation:** Iteratively integrating the ODEs or using discrete-event simulation for a finer representation. The Laplace transform remains valuable for partial analyses, such as local stability or small-disturbance behavior, but comprehensive system dynamics might be best captured by large-scale simulations.
3. **Hybrid Modeling:** In reality, many supply chains operate with both continuous and discrete elements (e.g., discrete shipping events, integer batch orders). A hybrid framework can embed continuous flows in certain segments while modeling discrete arrivals or reorder triggers in others. Laplace transforms can still handle the continuous portion, though specialized methods (including *Z-transforms* for discrete steps) might be needed for the entire system.

2.6.4.2 Policy Coordination and Global Optimization

By casting a multi-node supply chain as a system of coupled ODEs, managers gain a vantage point to coordinate policies across nodes. For instance, adjusting reorder thresholds at a central distribution center might shift the entire network's pole locations, thereby alleviating bullwhip-like oscillations further upstream. Similarly, centralizing certain forecasting or smoothing functions can act as a global filter, preventing local echelons from overreacting.

This perspective aligns well with modern *control theory* approaches that treat the supply chain as a closed-loop system with multiple feedback paths, each needing consistent tuning to achieve overall stability and efficiency.

2.6.5 LINKING TO NETWORK SCIENCE AND SUSTAINABILITY

In a forthcoming chapter on *Network Science in Logistics*, we will delve more deeply into how network topologies—ranging from simple lines or trees to complex, scale-free or small-world structures—affect logistics performance. Using tools from graph theory, we can measure connectivity (e.g., node degrees, clustering coefficients) and resilience (e.g., how the network withstands node or link failures). Integrating these measures with dynamic models helps us design supply chains that are not only stable in an operational sense but also robust against disruptions.

Sustainability goals add another layer of considerations: reverse logistics, circular economy flows, and emissions constraints. The same mathematical frameworks—ODEs, Laplace transforms, control policies—can be enriched to track resource usage, carbon footprints, or reusability cycles. The fundamental question remains how to manage multiple, interlinked nodes so that the entire supply network remains productive and resilient in an era of environmental challenges.

2.6.6 PRACTICAL GUIDELINES AND CONCLUDING REMARKS

To synthesize the main points:

1. **Scale with Caution:** Recognize that multi-node ODE systems grow rapidly in size and complexity. Decide whether a fully detailed, high-dimensional model is necessary or whether aggregated macro-models suffice.
2. **Identify Critical Flows and Nodes:** Not all nodes have equal impact on stability. Use partial analyses (e.g., local Laplace transform solutions) or network metrics to locate the "choke points" or "bottlenecks" most influential on system-wide behavior.
3. **Leverage Matrix Methods:** In linear or linearized systems, matrix formulations and inverse Laplace transforms provide global insights. Eigenvalue or pole analyses of the system matrix $\mathbf{A}$ or $(s\mathbf{I} - \mathbf{A})$ can reveal potential instability modes and guide how to tune local feedback gains.
4. **Combine Analytical and Numerical Approaches:** For large networks, partial fraction decomposition or purely symbolic inversion may be unwieldy. Simulation and numerical integration remain essential, but an

understanding of the underlying transforms clarifies where instabilities might originate and how to mitigate them.

5. **Plan for Robustness and Sustainability:** Coupling advanced network analysis with the dynamic models ensures that the logistics network can withstand disruptions and operate in an eco-efficient manner. Consider linking re-manufacturing or recycling nodes into the same mathematical framework to see how they impact overall flows.

In short, the mathematical techniques outlined in previous sections generalize to larger, more intricate supply chains. While the mechanics of taking the Laplace transform and solving ODEs remain fundamentally the same, the interpretation and management of system-level behavior demand a deeper understanding of both network structure and multi-echelon feedback. By integrating these dynamic models with concepts from network science and sustainable practices, logistics professionals can build robust, environmentally friendly supply chains for the challenges of the modern global economy.

3 Inventory Management as a Control System

3.1 INTRODUCTION

Inventory is among the most visible and critical assets in every logistics and supply chain operation. At a granular level, it represents the stock of items—from raw materials and components to work-in-process and finished goods—that a firm holds to meet demand, buffer against uncertainties, and optimize production schedules. Modern approaches in both industrial engineering and operations management increasingly treat inventory as a *control variable*, subject to systematic feedback loops that aim to maintain optimal levels. By viewing inventory decisions through the lens of control system analysis, we gain rigorous insights into mitigating the bullwhip effect, responding efficiently to demand fluctuations, and ensuring stable and cost-effective operation across the supply chain.

The foundation for this chapter builds upon the mathematical and system-theoretic viewpoints introduced in earlier parts of this book, where we examined ordinary differential equations (ODEs), Laplace transforms, and their utility for modeling logistic flows. We also explored how poles, zeros, and transfer functions provide a window into system stability and dynamic responsiveness. Now, we focus these tools explicitly on *inventory management*, illustrating how local reorder policies form feedback loops, how lead times introduce delays, and how carefully tuned gains or smoothing strategies can keep inventory within acceptable bounds in multi-echelon networks. The central objective is to equip you with the conceptual and analytical framework to design and analyze inventory control mechanisms in real-world supply chains.

Inventory theory is sometimes presented in a static or heuristic manner, where one calculates economic order quantities (EOQ) or safety stocks without deeply modeling the dynamic behavior of continuous inflows and outflows. A control-theoretic perspective reframes inventory as a continuously measured state whose target level is enforced by feedback actions that adjust production or order rates. This parallels physical control systems: just as a thermostat regulates temperature or a servo mechanism positions a robotic arm, a logistics manager regulates inventory levels by periodically observing stock and issuing orders. The mathematics of derivatives, transforms, poles, and gains appears naturally when exploring how a small deviation from the target leads to corrective action and how overshoot or delay might arise.

Open-Loop vs. Closed-Loop. If the supply chain reacts only to forecasted demand, inventory management may resemble an open-loop system: decisions are made solely on predicted usage, and there is limited or no corrective mechanism for real-time mismatches. Under closed-loop inventory control, feedback is harnessed.

DOI: 10.1201/9781003627906-3

The observed difference between current inventory $I(t)$ and a desired setpoint I_{target} triggers immediate changes in inflow, adjusting order quantities or production rates to minimize the discrepancy. This feedback principle helps to counter random shocks in demand and supply, although it can also amplify fluctuations if not carefully tuned.

Analogy to Physical Control Systems. Just as a motor speed controller senses rotational speed and applies voltage to reduce error, an inventory controller observes the difference between actual and desired stock and issues replenishment orders accordingly. The same mathematical tools—differential equations, Laplace transforms, stability criteria—become relevant. Casting inventory as a function $I(t)$ that is increased by inflows and decreased by outflows directly parallels standard control problems, but with the twist that *lead times* and *capacity constraints* can significantly influence the dynamics.

3.1.1 FEEDBACK LOOPS AND THE CORE INVENTORY EQUATION

A typical continuous-time inventory equation can be written as

$$\frac{dI}{dt} = \text{Inflow}(t) - \text{Outflow}(t).$$

If the outflow is driven by demand $D(t)$, and we decide inflow $Q(t)$ via a policy that compares $I(t)$ to a target I_{target}, we close a feedback loop. A simple proportional policy might define

$$Q(t) = k\left(I_{\text{target}} - I(t)\right),$$

where $k > 0$ is a gain. Substituting into the ODE gives

$$\frac{dI}{dt} = k\left(I_{\text{target}} - I(t)\right) - D(t).$$

We can take the Laplace transform to solve for $I(s)$ in the s-domain. This yields insights on how quickly inventory returns to its setpoint when disturbed by demand spikes, whether it overshoots, and whether there are frequency ranges of demand variations that cause large amplitude responses. The latter can be especially important if a business faces weekly or monthly demand cycles.

3.1.2 THE ROLE OF LEAD TIMES AND DELAYS

A critical factor in most inventory problems is the presence of lead times or shipment delays. If production or orders requested at time t arrive only at $t + \tau$, the feedback loop is no longer instantaneous. Instead of $Q(t)$ adding to $I(t)$, it is $Q(t - \tau)$ that arrives at the warehouse now, which can be modeled in the Laplace domain by the factor $e^{-s\tau}$. This time lag typically moves the system closer to potential instability if we hold gain k constant. Large delays demand more conservative policies or additional smoothing strategies; otherwise, the system can experience sustained oscillations.

Small mismatches in timing can cause the so-called *bullwhip effect*, where upstream echelons in the supply chain dramatically overreact to minor demand fluctuations, amplifying order variability. By extending the same differential equation approach and analyzing the resulting transfer functions, it becomes clear that longer lead times shift poles rightward, requiring lower feedback gains to avoid runaway or oscillatory behavior.

3.1.3 TRANSFER FUNCTIONS IN INVENTORY CONTROL

The transfer function captures how an input signal, such as demand $D(t)$, propagates through a system to an output variable, such as inventory $I(t)$. In a single-node setting with proportional feedback, we might define

$$G_{ID}(s) = \frac{I(s)}{D(s)},$$

the ratio of the Laplace transforms of inventory to demand. By identifying the poles of $G_{ID}(s)$, we can determine whether a step demand will settle quickly, induce an exponential decay, or produce oscillations. Zeros, meanwhile, can alter the phase or amplitude of the response, affecting phenomena like overshoot or undershoot.

When multiple echelons each have reorder policies, the supply chain can be viewed as a network of transfer functions. Outflow from one facility becomes an inflow to another, creating a cascade of transforms. Analyzing the chain in the s-domain helps to locate sources of potential amplification and to design consistent reorder strategies that minimize overall variance. This is particularly valuable if each node employs local forecasting or smoothing, which adds filter-like behaviors to the loop.

3.1.4 FEEDBACK POLICIES BEYOND SIMPLE PROPORTIONAL CONTROL

A proportional control policy $Q(t) = k\left(I_{\text{target}} - I(t)\right)$ is straightforward to implement, but more sophisticated options exist:

1. Integral components, where the order rate depends on the accumulated stock shortfall (or surplus) over time. This can eliminate persistent offsets but may create additional instability if lead times are large.
2. Derivative components, , which respond to the rate of change of inventory, potentially anticipating future imbalances. However, demand variability can render derivative control noisy.
3. Smoothing and filter-based policies, such as exponential smoothing, where the reorder rate depends partly on a filtered (rather than immediate) observation of demand or inventory discrepancy. This dampens high-frequency fluctuations at the expense of slower responsiveness to genuine shifts in demand.

Each of these controller enhancements modifies the transfer function. For instance, adding an integral term effectively introduces a pole at $s = 0$, which aids in driving steady-state errors to zero but risks magnifying lead-time-induced oscillations. On the other hand, smoothing imposes a low-pass filter on the demand signal, flattening rapid swings.

3.1.5 STABILITY AND THE BULLWHIP EFFECT

In control theory and dynamical systems, stability is characterized by the *location of poles* in the *complex s-plane*. This plane arises naturally when we take the Laplace transform of the system's differential equations, so that each solution or *mode* of the system corresponds to an exponential term e^{st}. Specifically, we write

$$s = \sigma + j\omega,$$

where σ and ω are real numbers and $j = \sqrt{-1}$. The real part $\Re(s) = \sigma$ determines whether the exponential term $e^{\sigma t}$ grows, decays, or remains constant over time:

1. If $\sigma < 0$, then $e^{\sigma t} \to 0$ as $t \to \infty$, so that mode decays.
2. If $\sigma > 0$, then $e^{\sigma t}$ grows unbounded as $t \to \infty$, making the system *unstable* in that mode.
3. If $\sigma = 0$, then the mode neither decays nor grows; it corresponds to neutral or marginal behavior (e.g., sinusoidal oscillations if $\omega \neq 0$).

Therefore, *stability in continuous time* requires all poles of the closed-loop transfer function to lie strictly in the *left half* of the complex s-plane (i.e., $\sigma < 0$). If any pole crosses the imaginary axis into $\sigma > 0$, small perturbations in that mode grow exponentially, causing unbounded inventory levels or self-reinforcing oscillations.

Even when the real parts of poles are close to $\sigma = 0$ but still negative, the system can exhibit *slowly decaying oscillations*, which may persist long enough to be disruptive in a practical supply chain context. The phenomenon is especially critical in *multi-echelon* environments. Even if each node is locally stable when treated in isolation, coupling delays and feedback from other nodes can shift the overall system's poles to the right, resulting in *global instability* or poorly damped oscillations.

Accordingly, in this chapter we investigate the *criteria for stability*: we examine how to choose or tune controller parameters—such as proportional gains k, integral gains in more advanced controllers, or smoothing constants in inventory replenishment rules—so that the *combined* system retains poles in the left half-plane. From a linear stability analysis perspective, one typically forms a *characteristic equation* of the form

$$\Delta(s) = 0,$$

where $\Delta(s)$ is a function that includes all the local feedback gains, lead times, and coupling terms from multiple nodes. We then solve for the roots s. If any such root has $\Re(s) \geq 0$, the supply chain is unstable or on the verge of instability.

Finally, we connect stability to the bullwhip effect by noting that aggressive feedback gains and nonzero lead times tend to make certain poles migrate toward the right half-plane or move farther from the imaginary axis, thereby producing *larger* amplitude swings in orders, inventories, and production. Even if the system remains technically stable ($\sigma < 0$), these poles can have small negative real parts with substantial imaginary parts, causing long-lived or pronounced oscillations. In a multi-echelon chain, such oscillations are amplified as they propagate upstream, inducing large variability in production schedules and costly disruptions, such as overtime labor, expedited transport, and excessive stockholding. Thus, understanding and managing the pole locations in the s-plane is key to preventing or at least mitigating the bullwhip effect.

3.1.6 MULTI-ECHELON DYNAMICS AND NETWORK EXTENSIONS

Real supply chains typically involve multiple echelons, each with unique lead times, demand patterns, and capacity constraints. Flows may converge or diverge. Reverse logistics channels can introduce closed loops of products returning for refurbishing or recycling. A system of coupled ODEs arises, where each node's inventory $I_i(t)$ depends on inflows from predecessor nodes and outflows to successor nodes. Feedback laws at each facility add further coupling.

This complexity is daunting, but the same fundamental principles hold: each node can be treated as a subsystem, with local or aggregated transfer functions. Analyzing the global system reveals potential resonances or bottlenecks. Inventory cannot be viewed solely from a local vantage; decisions in one echelon may reverberate through the entire network. Although deriving explicit analytic solutions for large networks can be challenging, partial fraction methods, numeric Laplace inverse algorithms, or direct time-domain simulations offer practical routes to understanding and mitigating bullwhip-like instabilities.

3.1.7 PRACTICAL CONCERNS IN IMPLEMENTING CONTROL-THEORETIC INVENTORY

One critical factor is data availability. Good control depends on accurate measurements of inventory, lead times, and demand. If on-hand stock is not tracked in real time or if demand signals are obscured by lost sales or backorders, the feedback loop might be operating on flawed information. Another concern is capacity constraints: if production or shipping capacity saturates, the linear control laws we model become nonlinear in practice, potentially degrading stability.

Discrete ordering intervals also matter. Many companies place orders weekly or monthly, transitioning the continuous ODE-based model to a discrete or hybrid system. While the underlying ideas of feedback, stability, and delays remain the same, the exact solution approach may shift to difference equations or Z-transforms.

3.1.8 ROADMAP OF THIS CHAPTER

This chapter lays out the foundational elements of *Inventory Management as a Control System* in several major steps:

1. **Mathematical Representations of Inventory Feedback Loops**. Introduces the fundamental equations and control diagrams that underpin the dynamic behavior of stock levels.
2. **Lead Times and Transient Dynamics in the Laplace Domain**. Examines how time delays, capacity constraints, and other factors enter into the

system's differential equations, affecting both transient and steady-state responses.

3. **Advanced Feedback Policies and Their Influence on Transfer Functions**. Explores more sophisticated reorder mechanisms (integral actions, smoothing filters) and discusses how each modifies the closed-loop system performance.
4. **Stability Analysis and the Bullwhip Effect**. Demonstrates why local reorder decisions can create significant oscillations throughout the supply chain if parameters are not carefully tuned.
5. **Practical Guidelines for Implementing Control-Theoretic Inventory Methods**. Concludes with actionable advice for real-world scenarios, addressing topics such as measurement accuracy, discrete time intervals, capacity constraints, and partial or uncertain lead times.

3.1.9 CONCLUDING REMARKS AND NEXT STEPS

By recasting inventory management in a control-theoretic framework, we gain a powerful toolset for understanding and improving supply chain performance. The same mathematics that govern control loops in engineering—differential equations, Laplace transforms, poles, and frequency responses—illuminate why overaggressive ordering can amplify small demand fluctuations, or why delays cause cyclical swings. Conversely, they also show how well-designed feedback can promote stable, low-variance inventory profiles.

The upcoming chapters build on these foundations. In Chapter 4, we will look at *Automatic Control in Logistics*, including the application of PID-like strategies to operational problems such as inventory replenishment and transport scheduling, bridging these ideas with classical Operations Research formulations. Chapter 5 delves into *Network Science in Logistics*, where we apply topological and connectivity analysis to supply networks and incorporate sustainability considerations. Chapter 6 explores *Feedback Loops and Kanban Systems*, focusing on Kanban's role as a dynamic signaling mechanism that enforces pull-based replenishment, tying these control concepts directly to lean manufacturing and reliability.

3.2 MATHEMATICAL REPRESENTATIONS OF INVENTORY FEEDBACK LOOPS

Inventory management can be viewed as a dynamic control system in which flows of material or product enter and leave a node based on decision rules, capacity constraints, and observed stock levels. By formulating these relationships mathematically, we obtain a set of equations and control diagrams that capture the essential feedback loops governing the behavior of inventory. This section examines how to represent these loops using ordinary differential equations (ODEs), transfer functions, and basic control diagrams, laying the groundwork for deeper analysis of stability, lead times, and advanced policies in subsequent sections.

3.2.1 CORE IDEA OF FEEDBACK IN INVENTORY

Feedback exists whenever a system's own state influences future decisions that alter that state. In inventory contexts, the deviation between an *actual* stock level and a *target* or *desired* level motivates reorder or production decisions. The primary variables typically include:

1. I(t), Inventory at time t,
2. D(t), Demand rate or outflow at time t,
3. Q(t), Inflow or reorder rate at time t.

An elementary inventory balance is:

$$\frac{dI(t)}{dt} = Q(t) - D(t),$$

in which $Q(t)$ increases inventory while $D(t)$ depletes it.

In a *feedback loop*, $Q(t)$ depends on the observed inventory $I(t)$ itself. One might define:

$$Q(t) = k\left(I_{\text{target}} - I(t)\right),$$

where $k > 0$ is a gain parameter that specifies how aggressively we respond to a shortage or surplus. High k means a strong correction for even small differences, while low k makes the system respond more gently.

3.2.2 BASIC ODE MODEL OF A SINGLE-NODE SYSTEM

Substituting the linear reorder policy into the inventory balance yields:

$$\frac{dI(t)}{dt} = k\left(I_{\text{target}} - I(t)\right) - D(t).$$

Rearranging:

$$\frac{dI(t)}{dt} + kI(t) = kI_{\text{target}} - D(t).$$

The left-hand side is reminiscent of a first-order linear ODE with a damping term $kI(t)$. Here, k also influences how rapidly inventory returns to the target after disruptions. The *particular solution* of the equation depends on $D(t)$, which can be constant, time-varying, or even impulsive. Solving in the Laplace domain often streamlines the analysis of transient and steady-state behaviors, revealing whether the system overshoots, oscillates, or converges smoothly to I_{target}.

3.2.3 CONTROL DIAGRAM INTERPRETATION

In control theory, as shown in Figure 3.1, we typically depict such a loop with block diagrams:

In such a diagram, I_{target} is added to the negative of $I(t)$, producing the error $e(t) = I_{\text{target}} - I(t)$. The *controller* block translates $e(t)$ into a command signal $Q(t)$.

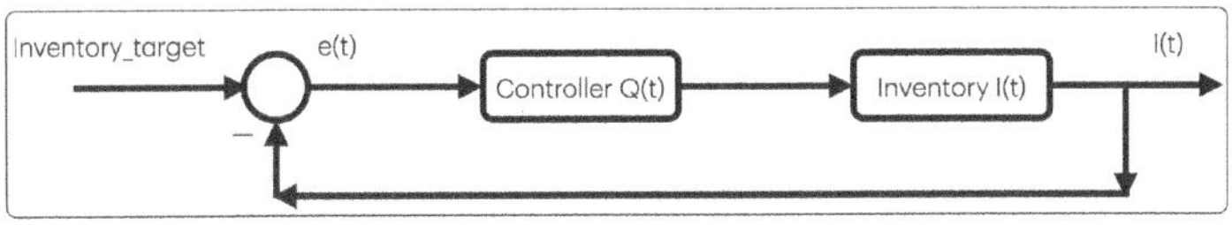

Figure 3.1 Block diagram.

The *plant* (or process) here is the physical inventory system, which integrates $Q(t)$ (inflow) and $D(t)$ (outflow). A sensor or measurement line closes the loop, sending $I(t)$ back for comparison against I_{target}.

This simple schematic can be extended to incorporate a time delay, τ, in the forward path if we have a known lead time, or to place additional logic or filtering in the controller block.

3.2.4 INCLUSION OF EXTERNAL DEMAND AND DISTURBANCE TERMS

Demand $D(t)$ often acts like a *disturbance* to the loop—an external signal that the controller strives to track or buffer against. In standard control diagrams, as shown in Figure 3.2, one might see $D(t)$ enter as an additive or subtractive term within the plant:

The integrator block $\int dt$ indicates that inventory accumulates or depletes over time. If $Q(t) > D(t)$, $I(t)$ grows; if $Q(t) < D(t)$, $I(t)$ shrinks. When $Q(t)$ is itself a function of $(I_{\text{target}} - I(t))$, the system is closed-loop. In scenarios with multiple inflows or concurrent demands, we may have additional summation nodes representing separate flows, each influenced by distinct policies or constraints.

3.2.5 DISCRETE VS. CONTINUOUS REPRESENTATIONS

Most of the equations so far assume continuous flows, which is a convenient abstraction for deriving differential equations and applying Laplace transforms. In practice, inventory changes often occur in discrete steps: shipments arrive in batches, orders are placed periodically, and time is segmented into intervals (e.g., days or weeks). One can still depict feedback loops with discrete or hybrid elements, replacing integrals by summations or using z-transforms instead of Laplace transforms.

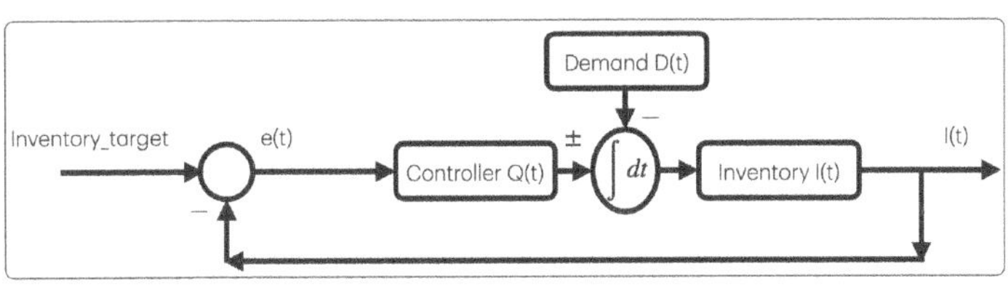

Figure 3.2 Inclusion of external disturbance.

The continuous-time perspective, however, remains powerful for capturing the *dynamic essence* of how inflows and outflows respond to state differences over time, and for highlighting the effect of lead time delays.

3.2.6 EXTENDING BEYOND SINGLE NODES

The diagrams and equations generalize to multiple interconnected facilities, each with its own stock variable $I_i(t)$ and local reorder policy. When Node A's output becomes Node B's input (with or without an additional transport delay), the blocks connect in series or in more complex network topologies (trees, hubs, or cycles). Partial fraction decomposition or state-space methods help to solve the resulting high-dimensional systems of ODEs. Even in a multi-echelon setting, the underlying theme persists: the relationship

$$\frac{dI_i}{dt} = \text{Inflow}_i(t) - \text{Outflow}_i(t),$$

paired with a feedback rule that determines inflow based on observed or forecasted states, underpins the overall dynamics.

3.2.7 KEY PARAMETERS IN A BASIC LINEAR INVENTORY MODEL

When translating real-world inventory decisions into a mathematical model, several parameters arise naturally. These parameters govern how quickly the system responds to discrepancies, how large the desired inventory level is how much time passes before orders arrive, and how demand patterns influence stock. Even though the equations may look relatively simple, the choice of parameter values can dramatically change the system's behavior from smooth and stable to oscillatory or even divergent. Below is a more detailed explanation of these parameters, illustrating why they are central to any inventory feedback loop.

1. **The Proportional Gain,** k
   In a proportional feedback system, the order rate $Q(t)$ is set to correct inventory discrepancies at a rate proportional to the difference between the current inventory $I(t)$ and the target level I_{target}. Concretely,

   $$Q(t) = k\left(I_{\text{target}} - I(t)\right),$$

   where k is the *proportional gain.* A large value of k means the system reacts vigorously to even slight differences in inventory: if $I(t)$ is just a bit below I_{target}, the reorder rate jumps sharply to compensate. This leads to fast corrections, which can be good for rapidly eliminating deficits or surpluses. However, if there is a time delay in receiving orders, a high k may cause the system to overcorrect and generate oscillations or instabilities. Conversely, a small k moderates the response, reducing the risk of wild swings but slowing the speed at which the system returns to the target.

2. **Desired Inventory Level, I_{target}**
The variable I_{target} represents the *setpoint* or goal for the inventory level. In classical control terms, it functions like the thermostat setting in a heating system: any deviation from I_{target} is considered an error that must be corrected by adjusting inflows. Changing I_{target} effectively shifts the system's equilibrium. For instance, raising I_{target} means you will aim to hold more stock on average, which can buffer against demand surges but increases holding costs. Lowering I_{target} reduces holding costs but may heighten the risk of stockouts if demand spikes. In mathematical models, I_{target} acts as a constant term in equations such as

$$\frac{dI}{dt} + kI(t) = kI_{\text{target}} - D(t),$$

dictating the steady-state inventory when demand $D(t)$ is constant.
3. **Time Delay or Lead Time, τ**
In practice, orders placed at time t do not instantly replenish inventory. There is often a lead time τ, which might include production, transportation, and administrative delays. In a continuous-time model, one way to represent this is to say that the inflow at time t depends on the order rate $Q(t-\tau)$, or equivalently, by multiplying the reorder signal in the s-domain by $e^{-s\tau}$.
A short lead time (τ close to zero) allows the system to correct inventory discrepancies almost immediately, making higher gains k feasible without risking large oscillations. On the other hand, a long lead time means feedback is outdated by the time it arrives, and if k is too large, the node may over-order or under-order significantly, leading to cyclical or even divergent behavior. As τ increases, the system's stability margin shrinks, so proper tuning of k and other parameters becomes crucial.
4. **Demand or Consumption Rate, $D(t)$**
The outflow $D(t)$ models how inventory is consumed or shipped out to meet external demand. In some scenarios, $D(t)$ might be constant (e.g., a steady demand for a raw material), but in most real-world cases it is time-varying and possibly random. Demand might exhibit daily or seasonal cycles, sudden spikes, or stochastic fluctuations.
The system's differential equation often takes the form

$$\frac{dI(t)}{dt} = Q(t-\tau) - D(t),$$

or a variant thereof, depending on whether or not we account explicitly for the time delay. Analyzing how $I(t)$ responds to various shapes of $D(t)$ (step changes, impulses, periodic patterns, or random noise) provides insight into how robust the control policy is. When $D(t)$ changes abruptly or unpredictably, a stable and well-tuned feedback system can adapt smoothly, whereas a poorly tuned system may exhibit large oscillations or persistent deviations from I_{target}.

When these parameters interact, the system can exhibit a range of behaviors:

1. **Smooth Convergence to I_{target}:** If k is chosen moderately and τ is not too large relative to the system's natural time scale, the inventory level tends to move smoothly toward I_{target} after a disturbance in demand or a temporary mismatch. The system is *overdamped* or *critically damped*, meaning that it does not oscillate significantly, and it settles to the desired level in a controlled fashion.
2. **Underdamped Oscillations:** If k is large and τ is moderate, the system may correct errors quickly but overcompensate, generating *underdamped* dynamics with visible oscillations around the target. These oscillations may decay over time if the real part of the poles remains negative, but they can drive up costs by requiring excess buffer stock or leading to short-lived stockouts.
3. **Sustained or Growing Divergences:** With high gain and significant lead times, the feedback loop might enter an *unstable* regime. The inventory can swing more and more wildly after each correction cycle, never settling to a steady state. In a real operation, this could manifest as repeated over-ordering followed by abrupt order stops, ballooning costs and risking severe shortages or excesses.

By experimenting with different values of k, I_{target}, τ, and the demand pattern $D(t)$, one can identify policy regimes that yield stable, efficient inventory control versus those prone to instability or protracted fluctuations. This understanding paves the way for advanced control strategies (e.g., integral or smoothing terms) and for careful coordination of multi-echelon networks, where multiple nodes each have their own gains and delays but must coexist harmoniously for the entire supply chain to function smoothly.

3.2.8 CONCLUSION

Mathematical representations of inventory feedback loops form the basis for analyzing and designing effective inventory management strategies. They transform intuitive notions of "reorder if we're below target" or "reduce orders if we're overshooting" into formal ODEs, block diagrams, and transfer functions. This precision reveals potential pitfalls (like overcorrection or bullwhip effects) and guides practical tuning of reorder rates, safety stocks, and lead-time offsets.

In the sections that follow, we build on these fundamental representations to incorporate specific delays, explore advanced feedback policies, and investigate how multiple interconnected facilities impact global stability. Each extension leverages the same underlying principle: treat inventory as a dynamic variable governed by inflows, outflows, and real-time feedback, ensuring that the system remains not just feasible or cost-effective, but dynamically robust in the face of demand variability and supply uncertainties.

3.3 LEAD TIMES AND TRANSIENT DYNAMICS IN THE LAPLACE DOMAIN

Supply chain lead times, transport delays, and capacity constraints introduce complexities that can drastically change both transient and steady-state behavior in inventory systems. The continuous-time modeling techniques introduced earlier, coupled with Laplace transforms, provide a powerful way to study these effects systematically. By incorporating delay terms (e.g., $e^{-s\tau}$) and representing real-world limits on production or distribution, one can anticipate how such factors might push a system toward oscillations, extended settling times, or even outright instability. This section explores how these elements are captured mathematically and how to interpret their impact on the dynamic response.

3.3.1 MODELING LEAD TIMES THROUGH TIME DELAY OPERATORS

In a simplified, delay-free model, the inflow $Q(t)$ directly appears in the inventory balance equation

$$\frac{dI(t)}{dt} = Q(t) - D(t).$$

However, real-life logistics typically involves a lead time τ, so the actual addition to inventory at time t depends on the order rate from *earlier* time $t - \tau$. A common approach is:

$$\frac{dI(t)}{dt} = Q(t-\tau)\,u(t-\tau) - D(t),$$

where $u(\cdot)$ is a unit step function ensuring that the inflow begins only after the delay. When we take the Laplace transform, we exploit the property

$$\mathscr{L}\{f(t-\tau)\,u(t-\tau)\} = e^{-\tau s}F(s),$$

where $F(s) = \mathscr{L}\{f(t)\}$. Therefore, a term $Q(t-\tau)$ in the time domain becomes $Q(s)\,e^{-\tau s}$ in the s-domain. This exponential factor $e^{-\tau s}$ often shifts the poles of the closed-loop system and can move them closer to the imaginary axis, reducing the stability margin.

3.3.2 TRANSIENT BEHAVIOR WITH DELAYS

Consider a first-order inventory loop with proportional control:

$$Q(t) = k\big(I_{\text{target}} - I(t)\big),$$

augmented by a fixed lead time τ. The state equation in the time domain might read

$$\frac{dI(t)}{dt} = k\big(I_{\text{target}} - I(t-\tau)\big)\,u(t-\tau) - D(t).$$

In the Laplace domain, $k\big(I_{\text{target}} - I(t-\tau)\big)$ becomes $k\left(\frac{I_{\text{target}}}{s} - I(s)\,e^{-\tau s}\right)$ once we handle constants and partial fractions carefully. If τ is negligible, the system's pole might be well in the left half of the s-plane for a moderate k. But as τ grows, the effective phase lag increases. Even modest gains can lead to underdamped or oscillatory transients, and large gains might push the system into near or actual instability.

3.3.3 CAPACITY CONSTRAINTS AND SATURATION

While lead time is a *temporal* delay, capacity constraints introduce a different form of delay-like effect. If production or transport capacity is capped at a finite value $Q_{\max}$, the order rate $Q(t)$ may saturate:

$$Q_{\text{actual}}(t) = \min\{k(I_{\text{target}} - I(t)), Q_{\max}\}.$$

Such saturation is a nonlinearity, and in the Laplace domain, it does not simply manifest as a neat exponential factor. Instead, one might approximate or linearize around an operating point, leading to piecewise models:

$$\frac{dI(t)}{dt} = \begin{cases} k(I_{\text{target}} - I(t-\tau)) - D(t), & \text{if } k(I_{\text{target}} - I(t-\tau)) \leq Q_{\max}, \\ Q_{\max} - D(t), & \text{otherwise.} \end{cases}$$

Though linear analysis must be applied with caution under these conditions, it can still offer insight into local stability. A system operating within capacity limits may behave linearly, but once capacity saturates, the dynamic signature changes—often causing delayed correction if demand spikes push the system into the constrained regime. This delayed return to normal operation can echo the effect of time lags and exacerbate oscillations.

3.3.4 LAPLACE TRANSFORM PERSPECTIVE ON DELAY AND CAPACITY

In the s-domain, a classical first-order system without delay has a transfer function

$$G(s) = \frac{k}{s+k}$$

(from Section 3.2, ignoring demand terms for the moment). Introducing a pure delay τ in the forward path modifies that transfer function to

$$\tilde{G}(s) = \frac{ke^{-\tau s}}{s+ke^{-\tau s}},$$

depending on how one structures the loop. While the exponential factor $e^{-\tau s}$ might appear straightforward, it fundamentally changes the pole structure. Poles are now solutions to $s + ke^{-\tau s} = 0$, a transcendental equation that yields an infinite number of possible solutions in the complex plane. Typically, only a few solutions lie close enough to the imaginary axis to dictate the system's slow or oscillatory modes, but those modes can be quite sensitive to changes in k or τ.

For capacity constraints, a *piecewise linear* or *linear-approximation* approach is often employed. When the system is not saturating, $Q(t)$ is unconstrained, and we can treat it as purely linear in the Laplace domain. Once $Q(t)$ attempts to exceed $Q_{\max}$, we switch to a different regime (a "clipped" or "sat" condition), for which a separate linear approximation can apply locally. This approach can be sufficient for local stability analysis, though more robust simulations or numerical methods might be necessary to capture large transients or repeated transitions in and out of saturation.

3.3.5 STEADY-STATE VS. TRANSIENT RESPONSES

Delays and constraints affect both how the system responds *over time* (transient dynamics) and what it ultimately settles to (steady state). Consider a demand step from D_1 to D_2. Without delay, a well-tuned feedback system may swiftly adjust production to D_2 and hold inventory near I_{target}. With a delay τ, the adjustment arrives late, possibly overshooting or undershooting the target inventory. The final steady-state inventory may still match I_{target} if there is enough capacity, but the route to get there could entail oscillatory swings. If the system frequently saturates at $Q_{\max}$, it might also stabilize at a new equilibrium below I_{target}, depending on the net inflow-outflow balance.

3.3.6 EXAMPLE: UNDER-DAMPED OSCILLATIONS WITH DELAY

A representative scenario involves a single node:

$$\frac{dI}{dt} = k\big(I_{\text{target}} - I(t-\tau)\big) - D(t),$$

with $D(t)$ constant for simplicity. A moderate k and small τ typically yield an exponential approach to the steady state. If τ grows or k is increased, the Laplace-domain equation can indicate complex conjugate poles with small negative real parts, implying sustained oscillations that slowly decay. As parameters push these poles closer to the imaginary axis, the system experiences larger and longer-lasting transients. If the real part shifts to zero or positive, the oscillations do not decay, effectively signaling instability in a real operation.

3.3.7 INSIGHTS FOR DESIGN AND CONTROL

From this analysis, several practical guidelines emerge:

1. **Delay Sensitivity.** Longer lead times exponentially amplify the risk of oscillations for a given feedback gain. Simple proportional control thus demands careful tuning: a gain k that was stable at $\tau = 1$ day might be far too large at $\tau = 7$ days.
2. **Buffering and Forecasting.** If large τ is unavoidable (e.g., international shipping), additional safety stock or more sophisticated forecasting can help mitigate wild swings. In the Laplace domain, partial smoothing of signals (adding a filter) can reduce high-frequency sensitivity but at the cost of slower response to genuine demand changes.

3. **Capacity Planning.** Sufficient capacity margin reduces the chance that $Q_{\text{actual}}(t)$ saturates during demand surges, preventing secondary delays in replenishing stock. If capacity is tight, local linear approximations can fail for large deviations, and a more cautious or multi-phase control strategy may be necessary.
4. **Stability vs. Responsiveness Trade-off.** Even if a small k avoids instability, it may yield sluggish correction of stockouts or surpluses. Real systems often choose a k that balances fast response against potential overshoot, leveraging knowledge of τ and capacity to maintain stable but sufficiently agile operation.

3.3.8 CONCLUSION

Time delays, capacity constraints, and other real-world complexities significantly alter how an inventory system behaves in both the transient and steady-state regimes. The Laplace transform framework systematically accommodates these factors, revealing how each enters the differential equations and changes the location of poles or modifies transfer functions. By carefully examining the interplay of delay and capacity, logistics managers and engineers can anticipate underdamped oscillations, extended settling times, or even instability—and they can design policies to avert these undesired dynamics. Subsequent sections will build on these results to explore advanced feedback mechanisms and multi-echelon structures, where lead times and constraints become even more pivotal in shaping system-wide performance.

3.4 ADVANCED FEEDBACK POLICIES AND THEIR INFLUENCE ON TRANSFER FUNCTIONS

The simplest forms of inventory control rely on proportional feedback, where the reorder rate is proportional to the gap between actual and desired stock. While this approach can work well for certain conditions, real-world supply chains often contend with volatile demand, significant lead times, and the risk of oscillations or steady-state offsets. To address these challenges, more sophisticated feedback policies incorporate integral actions, derivative elements, or smoothing filters. Each modification alters the closed-loop transfer function in specific ways, potentially improving system responsiveness, reducing persistent errors, or mitigating the bullwhip effect. This section outlines several advanced strategies, explains their underlying mathematics, and examines how they influence inventory dynamics in both transient and steady states.

3.4.1 INTEGRAL CONTROL FOR ELIMINATING STEADY-STATE ERRORS

A *proportional-integral* (PI) policy enhances the simple proportional law by adding an integral term to the reorder rate:

$$Q(t) = k_p\Big(I_{\text{target}} - I(t)\Big) + k_i \int_0^t \Big(I_{\text{target}} - I(\tau)\Big)\, d\tau.$$

The integral term accumulates the inventory deviation over time, steadily adjusting inflow if a mismatch persists. In classical control theory, this transforms the inventory feedback loop from a first-order system into one with an additional (often marginal) pole at $s = 0$, which enforces zero steady-state error in many demand scenarios (e.g., when $D(t)$ is constant).

PI control ensures that if demand or capacity shifts, the system will continue to correct for any inventory gap indefinitely, gradually driving the discrepancy to zero. Over the long run, this means inventory can settle exactly at I_{target}, whereas a purely proportional system might allow a small but nonzero offset.

If lead times are substantial or the gain k_i is too large, integral action can exacerbate oscillations. The integrator continuously ramps up the reorder signal even if the system is on the verge of overshooting. When combined with transport delays, this effect can drive underdamped or borderline unstable behavior. Careful tuning of k_p and k_i is therefore critical, typically employing methods analogous to those used in process control (e.g., root locus or frequency-domain analyses).

3.4.2 DERIVATIVE ELEMENTS AND PREDICTIVE CONTROL

A *proportional-derivative* (PD) or full *PID* policy can further incorporate a derivative term:

$$Q(t) = k_p\Big(I_{\text{target}} - I(t)\Big) + k_d \frac{d}{dt}\Big(I_{\text{target}} - I(t)\Big).$$

While this is less common in inventory management than in engineering processes, it can help anticipate the rate at which stock levels are changing. If inventory is dropping quickly, derivative control boosts the reorder rate preemptively, reducing lag-induced shortages.

Demand fluctuations or measurement noise can make a pure derivative term unstable or erratic, because the derivative amplifies high-frequency signals. If demand is highly unpredictable, the derivative component may oscillate or spike, complicating warehouse and transportation schedules. Hybrid or "pseudo-derivative" approaches often replace the raw derivative with a filtered version to dampen noise while still capturing predictive effects.

3.4.3 SMOOTHING FILTERS ON DEMAND OR REORDER SIGNALS

In many industrial settings, *demand smoothing* or *forecast filtering* is employed to reduce the sensitivity of reorder actions to abrupt changes. One popular method is exponential smoothing:

$$\hat{D}(t) = \alpha D(t) + (1-\alpha)\hat{D}(t-\Delta t),$$

where $0 < \alpha < 1$ and Δt is the update interval (which can be infinitesimal in a continuous approximation). The controller might then set

$$Q(t) = k\Big(I_{\text{target}} - I(t)\Big) + \hat{D}(t),$$

meaning the reorder signal depends on a *smoothed* demand estimate plus a feedback component. In the Laplace domain, smoothing often appears as a low-pass filter factor

$$F(s) = \frac{\alpha}{s+\alpha}$$

(matching discrete-time implementations approximately). By reducing high-frequency gain, the system dampens rapid oscillations. However, it also becomes slower to respond to genuine demand shifts, creating a trade-off between noise immunity and responsiveness.

3.4.4 IMPACT ON THE CLOSED-LOOP TRANSFER FUNCTION

Each advanced policy changes the shape of the closed-loop transfer function. For instance, a proportional controller yields a typical first-order term $\frac{k_p}{s+k_p}$ (in simplified form). Adding integral action places a factor $\frac{k_i}{s}$ in the loop, shifting the characteristic equation to

$$s\left(s+k_p\right) + k_i = 0,$$

when ignoring demand terms. This modifies the poles, potentially introducing an additional low-frequency mode that can eliminate steady-state offsets. Similarly, smoothing filters introduce additional polynomial or exponential factors that suppress certain frequency ranges in the demand signal, often acting as a *low-pass* filter in the s-domain.

Engineers analyzing these systems may plot Bode diagrams (gain and phase vs. frequency) or compute the system's poles and zeros. Smoothing generally pulls the frequency response down at higher frequencies, integral action boosts low-frequency gain, and derivative elements increase high-frequency slope. Understanding these transformations clarifies the trade-offs among stability margins, convergence time, and overshoot.

3.4.5 PRACTICAL CONSIDERATIONS IN LOGISTICS CONTEXTS

Although advanced feedback designs can substantially improve inventory management, a few practical points must be considered:

1. **Lead Time Sensitivity.** Integrators and derivative terms can magnify delay-induced lags. In systems with substantial τ, it is wise to keep integral gains modest or to incorporate lead time approximations in the controller design (e.g., Smith predictors).
2. **Measurement Noise and Demand Variability.** Derivative actions are especially prone to responding to noise spikes, and integral terms may chase random fluctuations, leading to overstock or understock cycles. Filters can help, but they also introduce a phase lag.
3. **Capacity Constraints.** Once any advanced control law tries to reorder at a rate exceeding capacity, the system effectively saturates and reverts to

piecewise dynamics outside the linear domain. Integrator wind-up is a well-known phenomenon in traditional process control, where the integral term accumulates a large error that continues to drive the system after the capacity limitation no longer applies. Techniques such as "anti-wind-up" logic can mitigate these challenges.

4. **Discrete Implementation.** Many supply chain operations place orders in daily or weekly intervals rather than continuously. While the principles of integral or smoothing control remain valid, they must be discretized carefully (e.g., via a Z-transform or difference equation approach) to avoid aliasing or unexpected stability issues.

3.4.6 EXAMPLE: COMBINED SMOOTHING AND INTEGRAL CONTROL

A notable approach combines demand smoothing with integral stock correction. The reorder rate might be:

$$Q(t) = \hat{D}(t) + k_i \int_0^t \Big(I_{\text{target}} - I(\tau)\Big) d\tau,$$

where $\hat{D}(t)$ is a filtered version of actual demand. In a stable regime, this ensures inventory eventually converges to I_{target} while preventing wild reactions to short-lived demand spikes. In the Laplace domain, it effectively places a filter on the demand input and adds a pure integration path on the inventory feedback branch. The poles and zeros shift accordingly, often yielding a second-order or higher system that must be tuned to balance responsiveness, stability, and noise rejection.

3.4.7 CONCLUSION

Advanced feedback policies expand beyond pure proportional control to tackle persistent offsets, mitigate demand noise, and account for future trends. Integral action systematically removes steady-state errors, derivative elements provide anticipatory corrections, and smoothing filters reduce high-frequency perturbations. Each policy shapes the closed-loop transfer function, shifting poles and altering how the system responds to demands or disturbances over time. When carefully tuned, these strategies can substantially enhance the resilience and performance of inventory management in dynamic, uncertain logistics environments. However, one must be mindful of lead times, capacity ceilings, and measurement noise, as these factors can undermine the theoretical benefits of advanced control if they are not accounted for in the design process. Subsequent sections explore how these ideas scale up to multi-echelon supply chains, where consistent or coordinated strategies at each node help maintain global stability and efficiency.

3.5 STABILITY ANALYSIS AND THE BULLWHIP EFFECT

A recurring challenge in inventory management is preventing local control decisions from causing wide-ranging disruptions throughout the supply chain. Even when each node (such as a warehouse, distribution center, or retailer) employs a seemingly reasonable reorder policy, subtle interactions can give rise to severe oscillations—an outcome famously known as the *bullwhip effect*. This section details why local decision rules, when combined with lead times and feedback gains, can destabilize multi-echelon networks. We also discuss the analytical techniques that reveal such instabilities and outline practical steps to mitigate them.

3.5.1 ORIGINS OF THE BULLWHIP EFFECT

The term "bullwhip effect" describes the phenomenon where small fluctuations in consumer demand propagate upstream in a supply chain, becoming increasingly larger at each echelon. Minor variations at the retail level might lead to disproportionately large swings in factory production or raw material orders. Several factors contribute:

1. **Lead Times and Information Delays:** Time lags separate the moment a facility realizes its inventory is too low from the moment replenishment arrives. If the reorder policy is aggressive, the node may overcorrect, not yet aware that a large delivery is already en route, causing an overshoot.
2. **Batch Ordering and Capacity Constraints:** When orders must be placed in discrete batches or subject to minimum order quantities, small changes in perceived demand can trigger big jumps in order size. If capacity saturates, the system may overcompensate once the constraint is lifted.
3. **Multiple Feedback Loops:** In a multi-echelon chain, each node uses local inventory signals to adjust inflows, effectively creating nested or cascaded loops. Demand spikes downstream trigger reorder surges upstream, leading to further surges, particularly if lead times or feedback gains are large.

The net result is that local logic can seem optimal from an individual node's viewpoint but collectively generate amplified oscillations that are costly in terms of inventory holding, production smoothing, and service level disruptions.

3.5.2 LINEAR STABILITY ANALYSIS AND POLE LOCATIONS

While the bullwhip effect is commonly illustrated with real-world anecdotes, one can analyze it rigorously using the continuous-time models and Laplace-domain techniques introduced earlier. In a simplified linear chain of n facilities, each node i has an inventory $I_i(t)$ governed by

$$\frac{dI_i(t)}{dt} = Q_i(t-\tau_i) - D_i(t),$$

where $Q_i(t)$ may be linked to $I_i(t)$ (or $I_{i-1}(t)$ in an upstream-downstream relationship) via a proportional or advanced feedback policy, and τ_i represents a lead time or delay for node i. Taking Laplace transforms yields a coupled set of algebraic equations, often forming a rational matrix

$$\mathbf{I}(s) = (\mathbf{A} - s\mathbf{I})^{-1}\mathbf{B}\mathbf{D}(s),$$

where $\mathbf{I}(s)$ and $\mathbf{D}(s)$ are vectors of transformed inventory and demand signals, and $\mathbf{A}$ encodes local reorder gains and delays. The poles of the global transfer function—i.e., the values of s that make $\det(\mathbf{A} - s\mathbf{I}) = 0$—govern stability. If any pole has a positive real part, the system is unstable: small demand perturbations are exponentially amplified over time.

In many multi-echelon or multi-node scenarios, a single echelon's reorder aggressiveness can push certain poles close to the imaginary axis. Even if those poles remain marginally in the left half-plane, the system may exhibit slow-decaying oscillations that practically cause substantial bullwhip effects. This results in undue variability in upstream orders, spikes in production capacity usage, and inevitable high inventory swings.

3.5.2.1 Conclusion

By linearizing the ODEs that describe each node's inventory and feedback policy, and by applying Laplace transforms, one obtains a matrix equation whose characteristic polynomial (including exponential terms for delays) determines the system's *poles*. Analyzing these poles makes it possible to *prove* when a system is stable or unstable, rather than relying on anecdotal evidence alone. This formalism explains how seemingly local choices of reorder gain k and lead times τ_i can drive the entire supply chain into large, persistent oscillations—the bullwhip effect. Armed with this insight, managers and engineers can then design safer feedback gains, reduce unnecessary lead times, or coordinate multi-echelon policies to ensure that the system's poles remain sufficiently in the left half of the complex plane, thereby promoting stable, damped responses to variability in demand.

3.5.3 ROOT LOCUS AND FREQUENCY RESPONSES

Classical control techniques such as *root locus* and *Bode plots* provide valuable tools for visualizing how changes in key parameters—like the reorder gain k or the lead time τ—affect the poles of a supply chain model. Understanding these effects is crucial for ensuring system stability and mitigating the bullwhip effect.

The *root locus* method plots the locations of the closed-loop poles in the complex s-plane as a key parameter, typically the reorder gain k, varies from 0 to ∞. The closed-loop poles are the roots of the characteristic equation:

$$1 + G(s) = 0,$$

where $G(s)$ is the open-loop transfer function of the system. For a simple supply chain model with proportional reorder gain, the open-loop transfer function might take the form:

$$G(s) = \frac{ke^{-s\tau}}{s + \alpha},$$

where k is the reorder gain, τ is the lead time, and α represents system-specific dynamics such as damping or additional control elements.

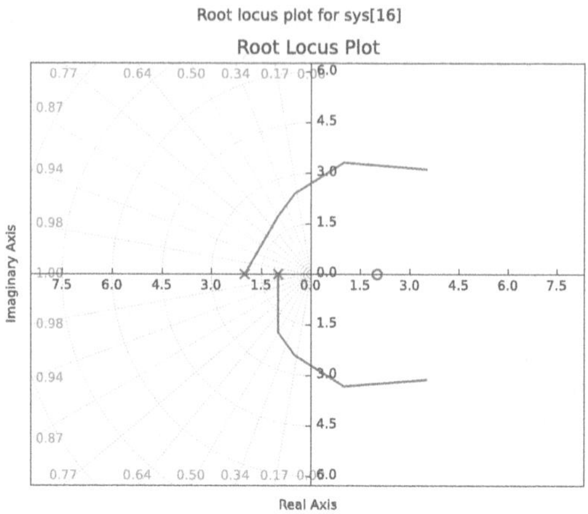

Figure 3.3 Root Locus Plot: Movement of closed-loop poles in the complex s-plane as the reorder gain k increases. The plot illustrates how poles shift from left to right, potentially crossing into the right half-plane as k grows, leading to instability.

As shown in Figure 3.3, as k increases, the root locus shows how the poles of the system move in the s-plane:

$$G(s) = \frac{ke^{-s\tau}}{s+\alpha}.$$

1. **Overdamped, Stable Behavior:** At moderate values of k, all poles lie well within the left half-plane ($\Re(s) < 0$), resulting in a stable and overdamped response with minimal oscillations.
2. **Approaching Instability:** As k increases, poles move toward the imaginary axis. If a pole crosses into the right half-plane ($\Re(s) > 0$), the system becomes unstable, exhibiting exponentially growing oscillations.
3. **Underdamped Oscillations:** Even before reaching instability, poles that approach the imaginary axis may result in underdamped or lightly damped oscillations, causing sustained or slowly decaying oscillatory behavior in inventory and order levels.
4. **Multi-Echelon Complexity:** In multi-echelon supply chains, the root locus becomes more complex due to the interactions between multiple nodes. Each additional echelon introduces more poles and potential interactions, making analytical root locus plots challenging. Numerical methods or simulation-based approaches are often necessary to accurately trace pole movements in such systems.

Bode plots offer another perspective by illustrating, as shown in Figure 3.4, how the magnitude and phase of the open-loop transfer function $G(j\omega)$ vary with

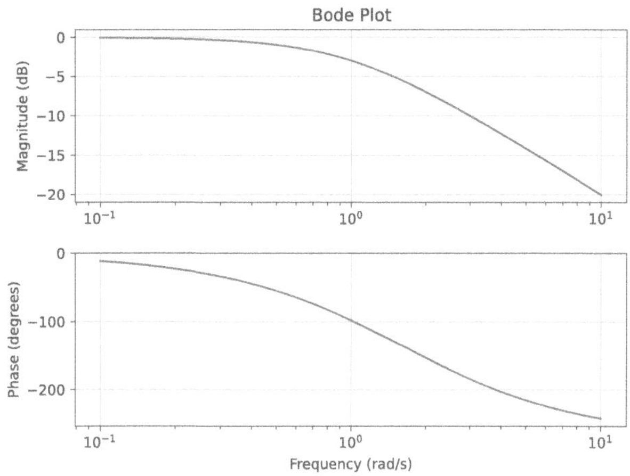

Figure 3.4 Bode Plot: Magnitude and phase of the open-loop transfer function $G(j\omega)$ versus frequency ω. Peaks in the magnitude plot indicate frequencies where the system amplifies demand fluctuations, contributing to the bullwhip effect. Phase lag introduced by lead time τ affects stability margins.

frequency ω. Consider the open-loop transfer function:

$$G(j\omega) = \frac{ke^{-j\omega\tau}}{j\omega + \alpha}.$$

1. **Magnitude Plot:** The magnitude $|G(j\omega)|$ indicates how much a demand fluctuation at frequency ω is amplified or attenuated by the supply chain. Peaks in the magnitude plot suggest that certain frequencies are being amplified, which can lead to significant variability upstream.
2. **Phase Plot:** The phase $\angle G(j\omega)$ shows the phase lag introduced by the system at each frequency. The lead time τ contributes a phase lag of $-\omega\tau$ radians. High phase lag can reduce the system's stability margin, making it more susceptible to oscillations.
3. **Gain and Phase Margins:** Stability criteria such as the *Gain Margin* and *Phase Margin* can be derived from the Bode plots. A low gain margin or phase margin indicates that the system is close to instability, meaning that small increases in gain k or delays τ could push the system into an unstable regime.
4. **Bullwhip Effect Connection:** If the Bode magnitude plot shows high gains at frequencies corresponding to typical demand cycles (e.g., weekly or monthly), these fluctuations are likely to be amplified as they propagate upstream, exacerbating the bullwhip effect.

Combining insights from root locus and frequency-domain analyses, we observe the following:

1. **Gain and Stability:** Increasing the reorder gain k can shift closed-loop poles to the right in the s-plane (root locus view), potentially crossing into the right half-plane and causing instability. Similarly, increasing lead time τ introduces additional phase lag, which can reduce phase margins and destabilize the system.
2. **Frequency Amplification:** High loop gains at specific frequencies (as shown in Bode magnitude plots) indicate that demand fluctuations at those frequencies are amplified rather than dampened. This resonance-like amplification leads to large swings in orders and inventories upstream, a hallmark of the bullwhip effect.
3. **Multi-Echelon Amplification:** In multi-echelon supply chains, the interactions between multiple nodes can compound the effects of individual pole shifts and frequency amplifications. Each additional echelon adds complexity to the root locus and Bode plots, making it easier for the system to exhibit amplified oscillations or instability if not properly controlled.
4. **Design Guidelines:** To mitigate the bullwhip effect, it is essential to:
   - Carefully tune the reorder gain k to ensure that all closed-loop poles remain in the left half-plane, providing adequate damping and avoiding excessive oscillations.
   - Minimize lead times τ where possible to reduce phase lag and improve phase margins.
   - Use control strategies that account for multi-echelon interactions, such as coordinated reorder policies or advanced smoothing techniques, to prevent local stability from leading to global instability.

3.5.4 PRACTICAL MANIFESTATIONS OF INSTABILITY

When a system is unstable or near-unstable, the following scenarios often arise:

1. **Excessive Inventory Oscillations:** Echelons see alternating periods of surplus and shortage. Warehouse shelves empty abruptly, prompting urgent orders that arrive too late, leading to subsequent overstock. The amplitude of these oscillations might grow if the pole real parts are close to zero or positive.
2. **Strained Production Schedules:** Upstream facilities endure bursts of manufacturing orders followed by lulls, making workforce and capacity planning difficult. Overtime costs rise, or machines remain idle during the lulls.
3. **Service-Level Deterioration:** Downstream nodes might experience frequent stockouts (due to under-ordering) or high expedite costs for emergency shipments if the system consistently lags behind real demand signals.
4. **Supplier Relationship Pressures:** Erratic orders degrade forecast accuracy, hamper supplier scheduling, and increase lead times further, creating a negative feedback loop that aggravates the bullwhip effect.

3.5.5 DAMPING THE BULLWHIP EFFECT

Although the bullwhip effect can be severe, several strategies can reduce it by effectively reshaping the pole locations and smoothing feedback responses:

1. **Moderate Feedback Gains:** Restrict overly aggressive reorder parameters (k values). This prevents large corrective orders in response to small inventory deviations, mitigating the risk that lead time delays will push the system toward oscillations.
2. **Demand Smoothing and Forecast Filtering:** As introduced previously, applying filters (e.g., exponential smoothing) to raw demand data prevents high-frequency fluctuations from dominating reorder signals. This effectively lowers the gain at frequencies prone to bullwhip amplification.
3. **Lead Time Reduction:** Shorter τ lessens the tendency to overreact with outdated information. Streamlined transportation, faster order processing, or local sourcing can significantly improve stability margins.
4. **Capacity Buffers:** Ample production or shipping capacity reduces the risk of saturating reorder rates. In the event of a spike, higher capacity can fulfill the surge without prolonged backlog or further time delays.
5. **Information Sharing:** Providing upstream nodes with actual consumer demand signals (rather than only a retailer's order signal) can reduce the nested feedback loops that amplify small variations. Collaborative forecasting or vendor-managed inventory schemes often curb bullwhip severity.
6. **Coordinated Policies:** Aligning reorder strategies across multiple echelons avoids the scenario where each node locally tunes a high feedback gain without considering upstream delays or constraints. Cross-functional teams or centralized controllers can ensure global damping rather than local optimization.

3.5.5.1 A Demonstration of Linear Stability Analysis in a Two-Echelon Chain

To illustrate how linear stability analysis can formally reveal the bullwhip effect, let us consider a simplified two-echelon supply chain. Suppose we have:

1. **Node 1 (Downstream/Retailer):** Tracks inventory $I_1(t)$, faces external demand $D_1(t)$, and places orders $Q_1(t)$ on Node 2.
2. **Node 2 (Upstream/Supplier):** Tracks inventory $I_2(t)$, faces internal demand $Q_1(t)$ from Node 1 plus any external demand $D_2(t)$, and places orders $Q_2(t)$ on some higher-level supplier (not explicitly modeled here).

For simplicity, we assume each node uses a *proportional feedback* on its own inventory plus a fixed lead time, and that external demands $D_1(t)$ and $D_2(t)$ are exogenous inputs. We aim to show how to form the linear system in the Laplace domain, find the characteristic equation (the "pole condition"), and see how aggressive reorder gains can push the system toward instability or underdamped oscillations.

3.5.5.2 Step 1: Inventory Balance Equations

Each node's inventory changes according to:

$$\frac{dI_1(t)}{dt} = Q_1(t-\tau_1) - D_1(t),$$

$$\frac{dI_2(t)}{dt} = Q_2(t-\tau_2) - \Big(Q_1(t) + D_2(t)\Big).$$

Here, τ_1 and τ_2 are lead times for Nodes 1 and 2, respectively, capturing the delay between placing an order and receiving it. (Node 2's outflow is the sum of internal demand from Node 1, $Q_1(t)$, plus any exogenous demand $D_2(t)$ that might also pull from Node 2's stock.)

3.5.5.3 Step 2: Local Proportional Reorder Policies

Let each node use a proportional law that attempts to keep its own inventory near a target $I_{i,\text{target}}$. Denote these reorder gains by k_1 and k_2:

$$Q_1(t) = k_1\,(I_{1,\text{target}} - I_1(t)), \qquad Q_2(t) = k_2\,(I_{2,\text{target}} - I_2(t)).$$

Although real supply chains may use more sophisticated or coordinated rules, this simple linear feedback is enough to demonstrate how local reorder decisions can generate or amplify oscillations, especially with nonzero lead times τ_1 and τ_2.

3.5.5.4 Step 3: Formulating the ODE System

Substituting $Q_1(t)$ and $Q_2(t)$ into the inventory balances gives:

$$\frac{dI_1(t)}{dt} = k_1\,(I_{1,\text{target}} - I_1(t-\tau_1)) - D_1(t),$$

$$\frac{dI_2(t)}{dt} = k_2\,(I_{2,\text{target}} - I_2(t-\tau_2)) - \Big(k_1\,(I_{1,\text{target}} - I_1(t)) + D_2(t)\Big).$$

These equations reflect how each node's inflow is dictated by its own backlog relative to I_{target}, but *delayed* by τ_i between ordering and receipt. Node 1's outflow is $D_1(t)$, while Node 2's outflow is $Q_1(t) + D_2(t)$.

For simplicity in analyzing stability (rather than performance under external demand), many demonstrations set $D_1(t) = D_2(t) = 0$ or treat them as small perturbations around a baseline demand. We focus on the homogeneous system to derive the characteristic equation. In practice, $D_i(t)$ can be reintroduced later as an exogenous forcing function that can reveal how the system's poles shape the amplitude of bullwhip effects.

3.5.5.5 Step 4: Taking Laplace Transforms

Let $I_1(s) = \mathscr{L}\{I_1(t)\}$ and $I_2(s) = \mathscr{L}\{I_2(t)\}$. We also denote constants in the Laplace domain:

$$\mathscr{L}\{I_{1,\text{target}}\} = \frac{I_{1,\text{target}}}{s}, \quad \mathscr{L}\{I_{2,\text{target}}\} = \frac{I_{2,\text{target}}}{s}.$$

Using the shifting property for delays, $\mathscr{L}\{f(t-\tau)u(t-\tau)\} = e^{-\tau s}F(s)$, and ignoring demand terms in the homogeneous model for clarity, we obtain:

$$sI_1(s) - I_1(0) = k_1\left(\frac{I_{1,\text{target}}}{s} - I_1(s)e^{-\tau_1 s}\right),$$

$$sI_2(s) - I_2(0) = k_2\left(\frac{I_{2,\text{target}}}{s} - I_2(s)e^{-\tau_2 s}\right) - k_1\left(\frac{I_{1,\text{target}}}{s} - I_1(s)\right).$$

In more compact form, these can be rearranged into a matrix equation of the form

$$\mathbf{A}(s)\begin{bmatrix} I_1(s) \\ I_2(s) \end{bmatrix} = \mathbf{C}(s),$$

where $\mathbf{A}(s)$ contains both polynomial and exponential-in-s (delay) terms, and $\mathbf{C}(s)$ encapsulates the initial conditions and target setpoints.

3.5.5.6 Expanding the Matrix Form of $\mathbf{A}(s)$

Below is an expanded discussion of how to rearrange the Laplace-domain equations into a matrix system, making the terms in $\mathbf{A}(s)$ explicit. This shows precisely how the exponential (delay) terms and the proportional gains k_i appear in the diagonal/off-diagonal entries.

From the homogeneous-system Laplace transforms in Step 4 (omitting external demands $D_i(t)$ for clarity), we had:

$$sI_1(s) - I_1(0) = k_1\left(\frac{I_{1,\text{target}}}{s} - I_1(s)e^{-\tau_1 s}\right),$$

$$sI_2(s) - I_2(0) = k_2\left(\frac{I_{2,\text{target}}}{s} - I_2(s)e^{-\tau_2 s}\right) - k_1\left(\frac{I_{1,\text{target}}}{s} - I_1(s)\right).$$

Our goal is to isolate $I_1(s)$ and $I_2(s)$ on the left side in a matrix multiply, producing something of the form

$$\mathbf{A}(s)\begin{bmatrix} I_1(s) \\ I_2(s) \end{bmatrix} = \mathbf{C}(s).$$

3.5.5.7 Rearranging the First Equation (Node 1)

Starting with

$$sI_1(s) - I_1(0) = k_1\left(\frac{I_{1,\text{target}}}{s} - I_1(s)e^{-\tau_1 s}\right),$$

we gather all the $I_1(s)$ terms on the left:

$$sI_1(s) + k_1 e^{-\tau_1 s} I_1(s) = k_1 \frac{I_{1,\text{target}}}{s} + I_1(0).$$

Factor out $I_1(s)$:

$$\left(s + k_1 e^{-\tau_1 s}\right) I_1(s) = k_1 \frac{I_{1,\text{target}}}{s} + I_1(0).$$

Hence, in the final matrix, the coefficient in front of $I_1(s)$ is

$$s + k_1 e^{-\tau_1 s},$$

and there is no $I_2(s)$ term in this first row (i.e. $a_{12}(s) = 0$).

3.5.5.8 Rearranging the Second Equation (Node 2)

From

$$sI_2(s) - I_2(0) = k_2\left(\frac{I_{2,\text{target}}}{s} - I_2(s)e^{-\tau_2 s}\right) - k_1\left(\frac{I_{1,\text{target}}}{s} - I_1(s)\right),$$

we expand and group like terms:

$$sI_2(s) - I_2(0) = k_2 \frac{I_{2,\text{target}}}{s} - k_2 I_2(s) e^{-\tau_2 s} - k_1 \frac{I_{1,\text{target}}}{s} + k_1 I_1(s).$$

Now gather all the $I_2(s)$ terms on the left:

$$sI_2(s) + k_2 e^{-\tau_2 s} I_2(s) - k_1 I_1(s) = k_2 \frac{I_{2,\text{target}}}{s} - k_1 \frac{I_{1,\text{target}}}{s} + I_2(0).$$

We can write this more compactly by factoring out $I_2(s)$:

$$\left(s + k_2 e^{-\tau_2 s}\right) I_2(s) - k_1 I_1(s) = k_2 \frac{I_{2,\text{target}}}{s} - k_1 \frac{I_{1,\text{target}}}{s} + I_2(0).$$

Notice this gives a coupling term $-k_1 I_1(s)$ on the left side (i.e. an off-diagonal entry in the matrix that multiplies $I_1(s)$).

3.5.5.9 Assembling $\mathbf{A}(s)$ and $\mathbf{C}(s)$

Putting both equations together, we get:

$$\underbrace{\begin{bmatrix} s + k_1 e^{-\tau_1 s} & 0 \\ -k_1 & s + k_2 e^{-\tau_2 s} \end{bmatrix}}_{\mathbf{A}(s)} \begin{bmatrix} I_1(s) \\ I_2(s) \end{bmatrix} = \underbrace{\begin{bmatrix} k_1 \frac{I_{1,\text{target}}}{s} + I_1(0) \\ k_2 \frac{I_{2,\text{target}}}{s} - k_1 \frac{I_{1,\text{target}}}{s} + I_2(0) \end{bmatrix}}_{\mathbf{C}(s)}.$$

Hence, in more explicit form:

$$\mathbf{A}(s) = \begin{bmatrix} s + k_1 e^{-\tau_1 s} & 0 \\ -k_1 & s + k_2 e^{-\tau_2 s} \end{bmatrix}, \quad \mathbf{C}(s) = \begin{bmatrix} k_1 \frac{I_{1,\text{target}}}{s} + I_1(0) \\ k_2 \frac{I_{2,\text{target}}}{s} - k_1 \frac{I_{1,\text{target}}}{s} + I_2(0) \end{bmatrix}.$$

3.5.5.10 Why $\mathbf{A}(s)$ Matters for Stability

From a control-systems perspective, the *characteristic equation* that determines stability or oscillatory modes can be found by setting $\det \mathbf{A}(s) = 0$. In effect, solving

$$\left(s + k_1 e^{-\tau_1 s}\right)\left(s + k_2 e^{-\tau_2 s}\right) = 0$$

leads to a transcendental equation in s. Because of the exponential terms $e^{-\tau_i s}$, this characteristic equation can admit complex-conjugate roots (or real roots on the positive side of the s-plane) if the gains k_1 and k_2 are large enough or the lead times τ_1 and τ_2 are sufficiently long. Such roots indicate *underdamped oscillations* or even outright instability, thus formally capturing the essence of the bullwhip effect in a linear-feedback model.

3.5.5.11 Step 5: Characteristic Equation and Poles

Setting initial conditions and target constants aside to focus on the homogeneous characteristic equation, we let $I_1(0) = I_2(0) = 0$ and $I_{1,\text{target}}, I_{2,\text{target}} \to 0$ for linear stability analysis. In that scenario, the system simplifies to finding nontrivial solutions of

$$\mathbf{A}(s)\begin{bmatrix} I_1(s) \\ I_2(s) \end{bmatrix} = \mathbf{0}.$$

A nontrivial solution in the Laplace domain exists only if

$$\det \mathbf{A}(s) = 0.$$

This determinant yields a transcendental *characteristic equation* because of the exponential factors $e^{-\tau_1 s}$ and $e^{-\tau_2 s}$, for example:

$$\det \mathbf{A}(s) = f\left(s, k_1, k_2, \tau_1, \tau_2\right) = 0.$$

The *poles* of the system are the values of s solving $f(s) = 0$. If *any* root has $\mathrm{Re}(s) > 0$, the system is unstable, meaning a tiny disturbance can grow exponentially over time. If poles lie on or very close to the imaginary axis (i.e., $\mathrm{Re}(s) = 0$ or near zero), the system exhibits persistent or slow-decaying oscillations, which in practice can produce large bullwhip swings.

3.5.5.12 Step 6: Demonstrating Instability via an Example Condition

As a more concrete illustration, imagine $\tau_1 = \tau_2 = \tau$ and $k_1 = k_2 = k$ for symmetry. A simplified version of the above system might collapse to a single scalar characteristic equation of the form

$$s^2 + \alpha s e^{-\tau s} + \beta e^{-\tau s} = 0,$$

where α, β depend on k, τ, and how I_1, I_2 couple. Numerical plots of s in the complex plane (or using a root-locus approach for k) show that beyond certain thresholds of k or τ, solutions migrate to the right half-plane. This confirms that *aggressive reorder*

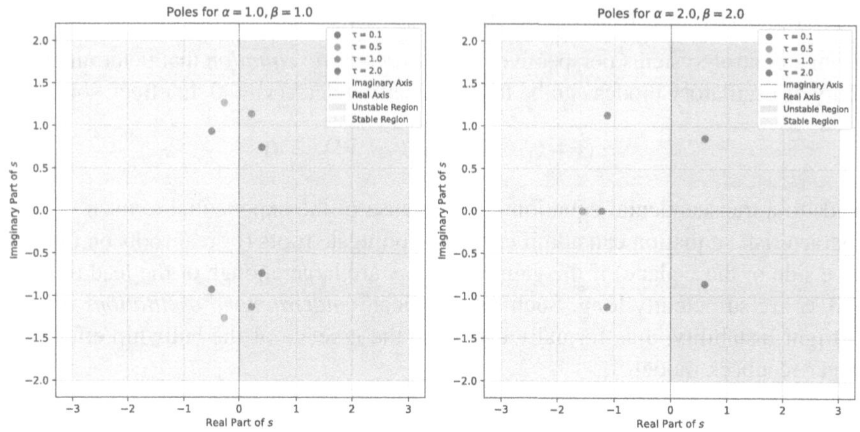

Figure 3.5 Poles of the system for α and β values. Stability and instability regions are shaded. Each marker corresponds to a different delay τ, showing how increasing delay impacts stability.

gains with *long lead times* can push the system into an unstable or near-unstable region, manifesting as bullwhip oscillations.

As shown in Figure 3.5 illustrates the poles of the system for different values of α and β, with varying delay τ. The stable region (left half-plane) and unstable region (right half-plane) are shaded in grayscale for clarity. The poles migrate into the unstable region as τ increases, highlighting the effect of delays on system stability. The trajectories of the poles depend on the parameters α and β, demonstrating the conditions under which the system becomes unstable.

3.5.5.13 Interpretation and Bullwhip Consequences

Even when $\mathrm{Re}(s) < 0$ but close to zero, the modes tied to those poles decay very slowly. In inventory terms, it may take many reorder cycles for the system to settle, and intermediate swings can be large. This is effectively a mild form of bullwhip: it does not explode over time, but it can impose costly transients.

If $f(s) = 0$ for some s with $\mathrm{Re}(s) > 0$, then any small mismatch or short-term demand spike grows exponentially, driving inventory away from equilibrium. Realistically, capacity saturation or managerial intervention often prevents truly unbounded inventory, but in practice it manifests as major oscillations or a recurring cycle of over-ordering followed by under-ordering.

With more than two echelons, the determinant $\det \mathbf{A}(s)$ becomes higher dimensional, typically containing multiple exponential delay terms. Each node's local reorder gain and lead time contributes to the overall dynamic. Certain parameter combinations push some poles rightward, leading to magnified upstream orders in response to small downstream signals. This is the hallmark of bullwhip in large-scale supply chains.

3.5.6 CONCLUSION

Stability analysis illuminates how well-intentioned, local reorder decisions—combined with lead times and multi-node feedback loops—can generate significant oscillations across the entire supply chain. The bullwhip effect is more than a theoretical curiosity; it manifests tangibly in excessive inventories, erratic production schedules, and deteriorating service levels. Mathematical modeling and Laplace-domain techniques make this effect visible, identifying the pole locations or frequency ranges that amplify small demand variations. Armed with this knowledge, supply chain managers and systems engineers can adopt cautious tuning of feedback gains, lead-time reduction measures, capacity planning, and synchronized strategies across echelons to counteract the bullwhip phenomenon. In the next section, we extend these principles to larger, multi-echelon or network-level configurations, where the interplay of local policies becomes even more complex and critical to system-wide performance.

3.6 PRACTICAL GUIDELINES FOR IMPLEMENTING CONTROL-THEORETIC INVENTORY METHODS

Control-theoretic approaches offer elegant insights into inventory dynamics, illuminating how feedback gains, lead times, and capacity constraints interact. Yet, translating these methods from theory to day-to-day logistics operations requires thoughtful adaptation. This section consolidates the main lessons into actionable guidance, acknowledging the realities of discrete ordering intervals, partial or uncertain lead times, measurement inaccuracies, and production or storage ceilings. By attending to these factors, managers and engineers can strike a robust balance between theoretical optimality and practical feasibility.

3.6.1 ADAPTING CONTINUOUS MODELS TO DISCRETE TIME INTERVALS

Although much of our analysis relies on *continuous-time* differential equations, most real supply chains place orders on discrete schedules (e.g., daily, weekly). A system that appears stable in the continuous limit can exhibit unexpected oscillations once discretized. For instance, an aggressive proportional gain that works with near-continuous updates may overshoot or even destabilize inventory levels if orders are only adjusted once a week.

When implementing a feedback law in discrete time, one often replaces the derivative or integral terms with finite-difference or summation approximations. A continuous proportional-integral (PI) law, for example, becomes

$$Q_t = k_p \left(I_{\text{target}} - I_t\right) + k_i \sum_{\tau=0}^{t} \left(I_{\text{target}} - I_\tau\right),$$

where t indexes discrete time periods. Although the broad dynamic principles remain the same, the “sampling” of feedback once per interval can introduce additional phase lag, requiring more conservative tuning of k_p and k_i compared to the continuous-time model.

Some practitioners adopt the *Z-transform*, a discrete analogue to the Laplace transform, to formalize control-law design. Poles in the z-domain must lie within the unit circle for stability (rather than in the left half-plane as in the continuous domain). For instance, a lead time of m discrete periods might appear as a factor z^{-m} in the z-domain. Ensuring stability translates into verifying that characteristic polynomials remain inside the complex unit circle, which parallels continuous-time pole analysis while adapting to the discrete update intervals.

3.6.2 DEALING WITH MEASUREMENT ACCURACY AND NOISE

A feedback loop performs only as well as the measurements it relies upon. If on-hand stock counts are systematically off (due to shrinkage, miscounting, or unrecorded returns), the reorder calculation may chase phantom deficits or surpluses. Introducing frequent cycle counts or RFID-based real-time tracking can reduce this noise, allowing the control system to target the true $I(t)$.

Even if inventory data are exact, demand signals are often noisy or intermittent. Smoothing filters (see Section 3.4) can prevent the system from responding excessively to short-lived spikes. However, each filter introduces phase delays. The more smoothing is applied, the slower the system recognizes genuine shifts in demand, risking stockouts if demand rises sharply.

In uncertain environments, engineers often prefer *robust tuning* that ensures acceptable performance under a range of measurement errors or parameter deviations, rather than optimizing for a single "best case." This may involve bounding the maximum reorder gain or applying a low-pass filter with a specific cutoff frequency to limit the impact of high-frequency noise.

3.6.3 CAPACITY CONSTRAINTS AND SATURATION

When reorder rates exceed production or transportation capacity, the system leaves the linear regime. Traditional pole-based analysis can underestimate the actual amplitude of oscillations or time delays that arise from queue build-ups. Approaches such as *anti-windup* for integral controllers or piecewise linear approximations provide partial solutions: the integral action is clamped so it cannot accumulate further once the capacity limit is reached, thereby reducing the risk of wild overshoot once the system exits the saturation zone.

Ample capacity buffer is the most straightforward way to avoid saturations during moderate demand surges. If capacity cannot be expanded, managers may hold additional safety stock to absorb demand spikes while the system is saturating. Although it raises holding costs, safety stock can stabilize the system by reducing the likelihood that actual inventory falls too low during capacity-induced delays.

3.6.4 PARTIAL OR UNCERTAIN LEAD TIMES

In practice, lead times may vary randomly (e.g., transportation delays, customs inspections). One approach is to model τ as a probability distribution, then analyze expected or worst-case outcomes. In a control-theoretic sense, uncertain τ can be approximated by a *mean lead time* plus a disturbance term. Although exact pole calculations become more intricate, scenario-based simulations can reveal if a chosen gain or smoothing filter remains stable for typical or maximum-likelihood lead times.

If lead times are uncertain, advanced techniques from control theory—like Kalman filters or extended state observers—can help estimate effective inventory in transit. By continuously updating one's best guess of how much product is en route and when it will arrive, the reorder policy can adapt more gracefully to fluctuations in τ.

3.6.5 MULTI-ECHELON COORDINATION AND POLICY ALIGNMENT

A node that tunes its gain k solely to minimize local stock variance may inadvertently push poles toward instability in the broader network. If the upstream node also has a large gain, lead times and cascade effects can amplify small disruptions. Coordinated policy design—where reorder gains, smoothing factors, or integral terms are chosen with a global objective—helps reduce bullwhip potential.

Sharing downstream demand data (rather than relying on immediate orders as the only signal) reduces the risk of repeated amplification. Direct visibility into consumer-level demand or point-of-sale information can smooth the entire chain, effectively lowering the required feedback gain at each echelon.

3.6.6 EMPIRICAL TUNING AND ITERATIVE DEPLOYMENT

Before rolling out a new control-based inventory policy chain-wide, managers can pilot it in a constrained environment (e.g., one region or product family) to gather data on real-time dynamics, capacity usage, and potential stockouts. Observed behaviors may deviate from purely linear predictions due to hidden nonlinearities (batch ordering, minimum order quantities, etc.). Iterative adjustments help refine the gain parameters and smoothing constants.

Continuous-time or discrete-event simulation packages allow one to import a proposed feedback design, incorporate actual lead-time distributions, capacity curves, and historical demand patterns, then check performance. Such "digital twins" of a logistics network can highlight where partial saturations or backlog might cause unexpected behavior. They also provide a safe environment to experiment with aggressive or integral-based policies.

An adaptive controller that periodically recalibrates parameters (like k, α in a smoothing filter, or integral gains) based on performance data can track evolving demand profiles or changing lead times. In a stable environment, this might be overkill, but in volatile markets, adaptive or model-predictive control can maintain better alignment between policy and reality.

3.6.7 SUMMARY OF KEY TAKEAWAYS

1. **Respect Discreteness:** Re-interpret continuous-time insights in a discrete-time framework, carefully checking that sampling intervals or batch ordering do not destabilize the loop.
2. **Measure Accurately:** Invest in reliable inventory and demand data, because even advanced controllers fail if their inputs are noisy or systematically biased.
3. **Anticipate Nonlinear Effects:** Plan for capacity saturation, batch constraints, and partial lead times. Adopt partial fraction approximations or piecewise logic where necessary, and consider anti-windup measures for integral policies.
4. **Coordinate Gains:** Link local reorder decisions to a broader multi-echelon strategy, limiting bullwhip amplification across the network.
5. **Simulate and Iterate:** Validate theoretical designs with numerical simulations or pilot deployments. Adjust parameters gradually to preserve stable operation.

3.6.8 CONCLUSION

Control-theoretic inventory methods hold substantial promise for maintaining stable and efficient stock levels in the face of fluctuating demand and lead times. Yet, the journey from an elegant differential equation to a workable real-world solution involves reconciling discrete order intervals, uncertain data, capacity ceilings, and partial lead-time knowledge. By implementing robust measurement protocols, carefully adjusting gain and filter parameters and iterating through simulations and pilot tests, one can capture the benefits of feedback control while avoiding the pitfalls of bullwhip oscillations and stockouts. The next chapters will continue to scale these concepts to complex networks, automatic control approaches (e.g., PID), and specialized environments such as Kanban systems, illustrating how each piece of the control-theoretic puzzle fits into broader logistics strategies.

4 Automatic Control in Logistics

4.1 INTRODUCTION

The efficient management of inventories and the seamless operation of supply chains are foundational to the competitiveness and sustainability of modern enterprises. As supply chains grow in complexity and scale, the need for sophisticated control mechanisms becomes increasingly paramount. Chapter 4, titled *Automatic Control in Logistics*, delves into the application of control theory principles—particularly Proportional-Integral-Derivative (PID) control—to logistical operations. By treating inventory replenishment as a dynamic feedback problem, we bridge the gap between theoretical control models and practical operations research strategies, equipping future logisticians with the tools to design responsive, resilient, and optimized supply chains.

Building upon the foundational concepts introduced in Chapter 3, where inventory was framed as a control variable subject to feedback loops, this chapter advances the discourse by integrating automatic control strategies. While earlier discussions focused on basic feedback mechanisms and their implications for system stability and the bullwhip effect, the current chapter elevates this understanding by introducing more nuanced control strategies that can handle real-world complexities such as variable demand, stochastic lead times, and capacity constraints.

4.1.1 THE ESSENCE OF AUTOMATIC CONTROL IN LOGISTICS

Automatic control systems are ubiquitous in engineering disciplines, from aerospace to manufacturing. In logistics, their application is equally transformative. The core idea is to autonomously regulate inventory levels, transportation flows, and production schedules to meet demand while minimizing costs and avoiding inefficiencies. This is achieved through the systematic use of sensors, actuators, and control algorithms that adjust operational parameters in real-time based on observed system states.

Consider a warehouse that maintains stock levels of various products. The objective is to ensure that inventory remains within desired thresholds to meet customer demand without incurring excessive holding costs or facing stockouts. Traditional inventory management might rely on periodic reviews and static reorder points, which can be suboptimal in dynamic environments. In contrast, an automatic control approach continuously monitors inventory levels and adjusts reorder quantities proactively, responding to fluctuations in demand and supply conditions.

DOI: 10.1201/9781003627906-4

4.1.2 PROPORTIONAL-INTEGRAL-DERIVATIVE (PID) CONTROL: THE WORKHORSE OF CONTROL THEORY

Among the plethora of control strategies, PID control stands out due to its simplicity, robustness, and effectiveness across a wide range of applications. A PID controller computes an error signal as the difference between a desired setpoint and a measured process variable (in our case, inventory level) and applies a corrective action based on proportional, integral, and derivative terms of this error.

Mathematically, the PID controller can be expressed as:

$$Q(t) = k_p e(t) + k_i \int_0^t e(\tau)\, d\tau + k_d \frac{de(t)}{dt},$$

where:

1. $Q(t)$ is the control action (e.g., reorder quantity),
2. $e(t) = I_{\text{target}} - I(t)$ is the error signal,
3. k_p is the proportional gain,
4. k_i is the integral gain, and
5. k_d is the derivative gain.

Each term plays a distinct role:

1. **Proportional** (k_p)**:** Provides an immediate response proportional to the current error. Higher k_p leads to faster corrections but can induce overshoot and oscillations.
2. **Integral** (k_i)**:** Accumulates past errors, eliminating steady-state offsets. However, excessive integral action can cause sluggishness or instability.
3. **Derivative** (k_d)**:** Predicts future error trends by considering the rate of change, thereby damping oscillations and improving system stability.

In the context of inventory management, a PID controller dynamically adjusts reorder quantities based on current inventory deviations, accumulated discrepancies, and anticipated inventory trends. This multifaceted approach enables more precise and stable inventory control compared to simple proportional or integral strategies.

4.1.3 INVENTORY REPLENISHMENT AS A FEEDBACK PROBLEM

Reiterating insights from Chapter 3, inventory replenishment can be viewed as a feedback control problem where the system aims to maintain inventory levels close to a desired setpoint. The fundamental balance equation is:

$$\frac{dI(t)}{dt} = Q(t) - D(t),$$

where $I(t)$ is the inventory level, $Q(t)$ is the inflow (reorder or production rate), and $D(t)$ is the outflow (demand rate).

In a PID-controlled system, $Q(t)$ is determined by the PID algorithm reacting to the error $e(t) = I_{\text{target}} - I(t)$. The interplay of proportional, integral, and derivative actions shapes how the system responds to disturbances (e.g., sudden demand spikes) and how it converges to the setpoint.

For instance, consider a scenario where demand suddenly increases. A proportional term immediately increases $Q(t)$ to counteract the deficit. The integral term ensures that any persistent error (due to delayed supply or ongoing high demand) is eventually corrected. The derivative term anticipates the trend, dampening potential overshoots by adjusting $Q(t)$ based on the rate at which $e(t)$ is changing.

4.1.4 MATHEMATICAL MODELING OF PID-CONTROLLED INVENTORY SYSTEMS

To rigorously analyze PID-controlled inventory systems, we employ differential equations and Laplace transforms. Consider the inventory balance equation augmented with a PID controller:

$$\frac{dI(t)}{dt} = Q(t) - D(t).$$

Substituting the PID control law:

$$Q(t) = k_p e(t) + k_i \int_0^t e(\tau)\, d\tau + k_d \frac{de(t)}{dt},$$

where $e(t) = I_{\text{target}} - I(t)$, we obtain:

$$\frac{dI(t)}{dt} = k_p(I_{\text{target}} - I(t)) + k_i \int_0^t (I_{\text{target}} - I(\tau))\, d\tau + k_d \frac{d}{dt}(I_{\text{target}} - I(t)) - D(t).$$

Simplifying, we have:

$$\frac{dI(t)}{dt} = k_p(I_{\text{target}} - I(t)) + k_i \int_0^t (I_{\text{target}} - I(\tau))\, d\tau - k_d \frac{dI(t)}{dt} - D(t).$$

Rearranging terms:

$$(1 + k_d)\frac{dI(t)}{dt} + k_p I(t) - k_i \int_0^t I(\tau)\, d\tau = k_p I_{\text{target}} - D(t).$$

Taking the Laplace transform (assuming zero initial conditions for simplicity):

$$(1 + k_d)sI(s) + k_p I(s) - \frac{k_i}{s} I(s) = \frac{k_p I_{\text{target}}}{s} - D(s).$$

Collecting like terms:

$$\left((1 + k_d)s + k_p - \frac{k_i}{s}\right) I(s) = \frac{k_p I_{\text{target}}}{s} - D(s).$$

Multiplying both sides by s to eliminate the denominator:

$$\left((1+k_d)s^2+k_p s-k_i\right)I(s)=k_p I_{\text{target}}-sD(s).$$

Thus, the transfer function from $D(s)$ to $I(s)$ is:

$$G_{ID}(s)=\frac{I(s)}{D(s)}=\frac{-s}{(1+k_d)s^2+k_p s-k_i}.$$

The denominator of this transfer function,

$$(1+k_d)s^2+k_p s-k_i=0,$$

is the *characteristic equation* whose roots (poles) determine the system's stability and transient behavior.

For the purposes of our analysis, we assume that all control parameters are real-valued. Specifically, the proportional gain k_p, integral gain k_i, and derivative gain k_d satisfy:

$$k_p,\quad k_i,\quad k_d\in\mathbb{R}.$$

This assumption ensures that the feedback actions are based on measurable, real-world quantities and facilitates the application of standard control-theoretic techniques to analyze system stability and performance.

To ensure that the real part of the roots is strictly smaller than zero ($\text{Re}(s)<0$), the parameters k_d, k_p, and k_i must satisfy one of the following four conditions:

1. $$1+k_d=0,\quad k_p=0,\quad k_i<0.$$

   This represents a degenerate case where the system simplifies, and stability depends solely on $k_i<0$.

2. $$1+k_d=0,\quad k_p>0,\quad x<-\frac{k_i}{k_p}.$$

   Here, the term $\text{Re}(s)$ must satisfy $\text{Re}(s)<-\frac{k_i}{k_p}$ to ensure stability.

3. $$1+k_d>0,\quad k_i>\frac{k_p^2}{4(1+k_d)}.$$

   In this case, the roots may be either real or complex, but the condition ensures that the real part remains strictly negative.

4. For $1+k_d>0$ and $k_i>0$, the imaginary part of the roots satisfies:

   $$\text{Im}(s)<\sqrt{\frac{k_i+x+(1+k_d)x^2}{1+k_d}},$$

   where $\text{Re}(s)$ must satisfy:

   $$\text{Re}(s)<-\frac{k_i}{k_p}.$$

The above four conditions collectively ensure that the roots of the characteristic equation have strictly negative real parts, guaranteeing system stability. These conditions account for both degenerate cases and cases with real or complex roots.

The nature of the roots is dictated by the discriminant:

$$\Delta = k_p^2 + 4(1 + k_d)k_i.$$

1. **Real and Distinct Roots ($\Delta > 0$):** The system has two real poles. Depending on the signs and magnitudes of the coefficients, the system can be overdamped (slow convergence without oscillations) or exhibit one stable and one unstable pole, leading to divergent behavior.
2. **Real and Repeated Roots ($\Delta = 0$):** The system is critically damped, achieving the fastest possible convergence to the setpoint without oscillations.
3. **Complex Conjugate Roots ($\Delta < 0$):** The system exhibits underdamped oscillatory behavior. The imaginary parts of the poles indicate the frequency of oscillations, while the real parts determine the rate of decay or growth.

1. **Proportional Gain (k_p):** Directly affects the damping and natural frequency of the system. Increasing k_p generally speeds up the response but can reduce damping, potentially leading to oscillations.
2. **Integral Gain (k_i):** Introduces a pole at the origin in the open-loop system, which can eliminate steady-state errors but also destabilizes the system by adding a negative feedback loop that can interact adversely with delays and high k_p values.
3. **Derivative Gain (k_d):** Adds a predictive element that increases system damping, helping to counteract the oscillatory tendencies introduced by high k_p and k_i. However, excessive k_d can make the system overly sensitive to noise and lead to erratic control actions.

4.1.5 IMPLICATIONS FOR OPERATIONS RESEARCH

Operations Research (OR) is inherently concerned with optimizing complex systems and decision-making processes. The integration of control theory into logistics enriches OR methodologies by introducing dynamic, real-time feedback mechanisms that can adapt to changing conditions. Traditional OR models often assume static or pre-planned actions, whereas control-theoretic approaches embrace the system's dynamism, allowing for continuous adjustment based on observed states.

The mathematical rigor of control theory complements the optimization focus of OR by providing tools to analyze system stability, responsiveness, and robustness. For instance, when designing inventory policies, OR practitioners can use PID control models to ensure that solutions are not only cost-effective but also dynamically resilient against uncertainties and perturbations.

Furthermore, the use of Laplace transforms and frequency-domain analysis enables the quantification of how control policies respond to various demand patterns, facilitating the design of policies that minimize both costs and variability. This synergy between control theory and OR fosters more holistic and adaptive supply chain strategies, capable of thriving in volatile and uncertain environments.

4.1.6 OVERVIEW OF CHAPTER STRUCTURE

This chapter is organized to guide you through the application of automatic control principles to logistics operations systematically. The progression is as follows:

1. **Fundamentals of PID Control in Logistics:** We begin by revisiting the PID control architecture, contextualizing its components within inventory management scenarios. Through mathematical derivations, we illustrate how each PID term influences the inventory dynamics and system response.
2. **Incorporating Lead Times and Delays:** Building on the earlier discussions of lead times, we examine how delays are incorporated into PID-controlled systems. We demonstrate how delays impact the transfer function and stability, and discuss strategies to mitigate adverse effects through controller adjustments or additional filtering.
3. **Advanced Control Strategies: Beyond PID:** While PID controllers are powerful, certain logistical challenges necessitate more advanced strategies. We introduce concepts such as model predictive control (MPC), adaptive control, and fuzzy logic controllers, comparing their advantages and applicability to different logistics contexts.

Each section is meticulously designed to blend mathematical rigor with practical insights, ensuring that you not only understand the underlying principles but also appreciate their real-world relevance and applicability.

4.1.7 CONCLUDING REMARKS

Automatic control in logistics transcends traditional inventory management by introducing a dynamic, responsive framework capable of adapting to real-time fluctuations and uncertainties. By applying PID control and other advanced control strategies, logisticians can design systems that not only meet current demand efficiently but also anticipate and mitigate future disruptions.

This chapter serves as a critical juncture in your journey through the mathematical underpinnings of logistics. It bridges the gap between static inventory theories and dynamic, real-time control systems, equipping you with the knowledge and tools to innovate and optimize supply chains in an increasingly complex and volatile global marketplace.

As you engage with the subsequent sections, maintain a focus on both the theoretical constructs and their practical implementations. The interplay between mathematics and logistics operations is intricate, but mastering this synergy is essential for advancing in the field of supply chain management and operations research.

4.2 FUNDAMENTALS OF PID CONTROL IN LOGISTICS

Proportional-Integral-Derivative (PID) control is a foundational strategy in automatic control systems, lauded for its simplicity and effectiveness across diverse applications. In the context of logistics, particularly inventory management, PID controllers offer a structured approach to dynamically adjusting reorder quantities to maintain inventory levels close to desired targets. This section delves deeper into the PID control architecture, contextualizing its components within inventory management scenarios. Through rigorous mathematical derivations, we elucidate how each PID term—proportional, integral, and derivative—affects inventory dynamics and system response.

4.2.1 REVISITING PID CONTROL ARCHITECTURE

A PID controller synthesizes three distinct control actions—proportional, integral, and derivative—each addressing different aspects of the error signal $e(t) = I_{\text{target}} - I(t)$, where $I(t)$ denotes the current inventory level and I_{target} is the desired inventory setpoint. The control action $Q(t)$ (reorder quantity or production rate) is computed as:

$$Q(t) = k_p e(t) + k_i \int_0^t e(\tau)\,d\tau + k_d \frac{de(t)}{dt},$$

where:

- k_p is the *proportional gain*,
- k_i is the *integral gain*, and
- k_d is the *derivative gain*.

Each term serves a unique purpose in the control strategy:

1. **Proportional Term ($k_p e(t)$):** Provides an immediate response proportional to the current error, facilitating swift corrective actions.
2. **Integral Term ($k_i \int_0^t e(\tau)\,d\tau$):** Accumulates past errors, ensuring that persistent discrepancies are eradicated by integrating the error over time.
3. **Derivative Term ($k_d \frac{de(t)}{dt}$):** Anticipates future errors by responding to the rate of change of the error, thereby damping oscillations and enhancing system stability.

4.2.2 CONTEXTUALIZING PID IN INVENTORY MANAGEMENT

In inventory management, maintaining optimal stock levels is crucial for balancing holding costs against service level requirements. Traditional methods, such as Economic Order Quantity (EOQ) or fixed reorder points, often fail to adapt dynamically to fluctuating demand and supply conditions. PID control introduces a responsive mechanism that continuously adjusts reorder quantities based on real-time inventory deviations, ensuring a more agile and stable inventory system.

Scenario Illustration: Consider a warehouse managing the inventory of a high-demand product. Sudden spikes in customer orders can deplete inventory rapidly, risking stockouts and lost sales. Conversely, periods of low demand can lead to excessive inventory, increasing holding costs and potential obsolescence. A PID-controlled reorder policy dynamically responds to these fluctuations:

1. The *proportional* term reacts immediately to inventory shortages or surpluses.
2. The *integral* term corrects for any persistent discrepancies, such as consistent underordering during high demand periods.
3. The *derivative* term dampens the response to prevent overshooting the target inventory level, ensuring smooth adjustments.

4.2.3 SYSTEM RESPONSE ANALYSIS

To visualize how PID parameters influence system behavior, consider the step response of the system to a sudden increase in demand $D(t)$. The step response characterizes how the inventory level $I(t)$ adjusts over time to reach the new equilibrium.

When the system is overdamped, the response to a step demand is slow but smooth, with no oscillations. This occurs when the proportional and integral gains are set to values that provide sufficient damping relative to the system's natural frequency.

An underdamped system exhibits oscillatory behavior around the setpoint. This can be desirable to some extent, as it allows the system to respond quickly, but excessive oscillations can lead to inefficiencies such as overstocking or stockouts.

A critically damped system strikes the optimal balance between speed and stability, reaching the setpoint as quickly as possible without oscillating. Achieving critical damping requires precise tuning of the PID parameters.

If the PID parameters are set too aggressively, particularly k_p and k_i, the system can become unstable. This results in growing oscillations or divergence from the setpoint, necessitating corrective measures such as reducing gains or introducing additional damping.

4.2.4 PRACTICAL EXAMPLE: TUNING PID PARAMETERS FOR INVENTORY CONTROL

Consider a warehouse aiming to maintain an inventory level $I_{\text{target}} = 100$ units. Suppose the current inventory level $I(t)$ drops below the target due to increased demand. The warehouse employs a PID controller to adjust the reorder quantity $Q(t)$.

Given Parameters:

$$k_p = 2, \quad k_i = 1, \quad k_d = 0.5.$$

$$\text{Characteristic Equation:} \quad (1+0.5)s^2 + 2s - 1 = 0 \quad \Rightarrow \quad 1.5s^2 + 2s - 1 = 0.$$

$$\text{Solutions:} \quad s = \frac{-2 \pm \sqrt{4+6}}{3} = \frac{-2 \pm \sqrt{10}}{3} \approx 0.3874, \quad -1.0541.$$

Interpretation:

1. One pole in the right half-plane ($s \approx 0.3874$): Indicates an unstable mode.
2. One pole in the left half-plane ($s \approx -1.0541$): Represents a stable, decaying mode.

Outcome: The presence of a pole with a positive real part signifies that any small deviation from the target inventory will grow exponentially over time, leading to unstable and oscillatory inventory levels.

To stabilize the system, the PID parameters must be tuned to ensure all poles lie in the left half-plane. One approach is to reduce the proportional and integral gains:

$$k_p = 1.5, \quad k_i = 0.5, \quad k_d = 0.5.$$

$$\text{Characteristic Equation:} \quad (1+0.5)s^2 + 1.5s - 0.5 = 0 \quad \Rightarrow \quad 1.5s^2 + 1.5s - 0.5 = 0.$$

$$\text{Solutions:} \quad s = \frac{-1.5 \pm \sqrt{2.25+3}}{3} = \frac{-1.5 \pm \sqrt{5.25}}{3} \approx \frac{-1.5 \pm 2.2913}{3}.$$

$$s_1 \approx 0.2638, \quad s_2 \approx -1.5976.$$

Interpretation: Still, one pole remains in the right half-plane, indicating ongoing instability. Further reduction of gains is necessary:

$$k_p = 1.0, \quad k_i = 0.4, \quad k_d = 0.5.$$

$$\text{Characteristic Equation:} \quad 1.5s^2 + 1.0s - 0.4 = 0.$$

$$\text{Solutions:} \quad s = \frac{-1.0 \pm \sqrt{1.0+2.4}}{3} = \frac{-1.0 \pm \sqrt{3.4}}{3} \approx \frac{-1.0 \pm 1.8439}{3}.$$

$$s_1 \approx 0.2813, \quad s_2 \approx -1.2813.$$

Outcome: The system remains unstable. To achieve stability, it may be necessary to either further decrease k_p and k_i or introduce more derivative action to enhance damping.

4.2.5 PID TUNING METHODS

Effective PID control hinges on the appropriate selection of k_p, k_i, and k_d. Several methodologies exist for tuning these parameters:

A heuristic tuning method where k_p is increased until the system exhibits sustained oscillations (ultimate gain K_u) with a corresponding oscillation period P_u. PID parameters are then set based on empirical formulas derived from K_u and P_u.

Another empirical approach that considers the system's open-loop response to step inputs, providing formulas to calculate PID parameters that optimize both stability and responsiveness.

Employ mathematical optimization techniques to minimize a cost function, such as the Integral of Time-weighted Absolute Error (ITAE) or the Integral of Squared Error (ISE), subject to stability constraints.

Modern control systems often leverage software tools that simulate system responses and iteratively adjust PID parameters to achieve desired performance metrics.

4.2.6 PRACTICAL CONSIDERATIONS IN PID IMPLEMENTATION

While PID controllers are powerful, their real-world implementation in logistics demands attention to several practical factors:

Accurate and timely inventory data are critical. Delays or inaccuracies in measurement can lead to erroneous control actions, exacerbating inventory swings. Implementing robust data acquisition systems and real-time tracking technologies (e.g., RFID) can enhance measurement fidelity.

In practice, orders are placed at discrete intervals (e.g., weekly or monthly). Discretizing the PID controller requires careful consideration to avoid introducing additional phase lag or instability. Techniques such as discretization via the Z-transform or employing discrete-time control laws can mitigate these issues.

Physical limitations on production or storage capacities mean that reorder quantities cannot be adjusted infinitely. Incorporating saturation limits into the PID controller prevents attempts to order beyond capacity, but also necessitates mechanisms like anti-windup to handle integral term constraints.

Demand signals are often noisy or subject to sudden, unpredictable changes. The derivative term can amplify high-frequency noise, leading to erratic control actions. Implementing filters or limiting the derivative action can help maintain system stability.

Supply chains are dynamic, with varying lead times, demand patterns, and operational conditions. Adaptive PID controllers that can adjust their parameters in response to changing environments enhance robustness and maintain optimal performance over time.

Seamless integration of PID controllers with existing inventory management systems, Enterprise Resource Planning (ERP) software and other logistical tools is essential for effective implementation. Ensuring interoperability and data consistency across systems facilitate smoother control actions and better overall supply chain performance.

4.2.7 EXAMPLE: DESIGNING A PID CONTROLLER FOR INVENTORY REPLENISHMENT USING THE ZIEGLER-NICHOLS METHOD

To illustrate the practical application of PID control in logistics, we will design a PID controller for an inventory replenishment system using the *Ziegler-Nichols* tuning method. This example demonstrates how to systematically determine the proportional (k_p), integral (k_i), and derivative (k_d) gains to achieve a balanced and stable inventory system.

Consider a warehouse managing the inventory of a high-demand product. The goal is to maintain the inventory level close to a desired setpoint while minimizing holding costs and avoiding stockouts. The key parameters and objectives of the system are as follows:

1. **Setpoint:** $I_{\text{target}} = 200$ units.
2. **Initial Inventory:** $I(0) = 150$ units.
3. **Demand Rate:** $D(t) = 50$ units/day.

4. **Lead Time:** $\tau = 2$ days.
5. **Objective:** Maintain inventory close to I_{target} while minimizing holding and shortage costs.

4.2.7.1 Step 1: Understanding the Ziegler-Nichols Method

The Ziegler-Nichols (Z-N) method is a heuristic approach to tune PID controllers based on the system's response to certain tests. There are two primary Z-N tuning methods:

1. **Step Response Method:** Utilizes the system's response to a step input to determine tuning parameters.
2. **Oscillation Method (Ultimate Gain Method):** Involves increasing the proportional gain until the system exhibits sustained oscillations, then uses the corresponding gain and oscillation period to calculate PID parameters.

In this example, we will employ the *Oscillation Method* due to its effectiveness in providing a balance between responsiveness and stability.

4.2.7.2 Step 2: Conducting the Oscillation Test

Objective: Determine the *ultimate gain* (K_u) and the *ultimate period* (P_u).

1. **Set Up the Controller:** Begin with the integral and derivative gains set to zero ($k_i = 0$, $k_d = 0$). The controller reduces to a simple proportional controller:
$$Q(t) = k_p e(t) = k_p \left(I_{\text{target}} - I(t)\right).$$
2. **Increase Proportional Gain (k_p):** Gradually increase k_p from zero until the system exhibits *sustained oscillations*—a state where the inventory level oscillates with constant amplitude without damping or growing over time.
3. **Record K_u and P_u:** Once sustained oscillations are achieved, note the corresponding proportional gain as the *ultimate gain* (K_u) and the period of oscillation as the *ultimate period* (P_u).

4.2.7.3 Step 3: Applying Ziegler-Nichols Tuning Rules

The Z-N method provides empirical formulas to calculate the PID parameters based on K_u and P_u. The recommended settings for different controller types are as follows:

Controller Type	k_p	k_i	k_d
P	$0.5K_u$	-	-
PI	$0.45K_u$	$\frac{1.2K_u}{P_u}$	-
PID	$0.6K_u$	$\frac{2K_u}{P_u}$	$\frac{K_u P_u}{8}$

For this example, we will focus on the *PID* controller.

4.2.7.4 Step 4: Example Calculation

Assume through experimentation, we determine:

$$K_u = 3, \quad P_u = 5 \text{ days}.$$

Applying the Ziegler-Nichols formulas for a PID controller:

$$k_p = 0.6 \times K_u = 0.6 \times 3 = 1.8,$$

$$k_i = \frac{2 \times K_u}{P_u} = \frac{2 \times 3}{5} = 1.2,$$

$$k_d = \frac{K_u \times P_u}{8} = \frac{3 \times 5}{8} = 1.875.$$

Thus, the initial PID parameters are:

$$k_p = 1.8, \quad k_i = 1.2, \quad k_d = 1.875.$$

4.2.7.5 Step 5: Simulating the PID-Controlled System

With the determined PID parameters, we simulate the system's response to a step increase in demand. The inventory balance equation incorporating the PID controller is:

$$\frac{dI(t)}{dt} = k_p(I_{\text{target}} - I(t)) + k_i \int_0^t (I_{\text{target}} - I(\tau))\,d\tau + k_d \frac{d}{dt}(I_{\text{target}} - I(t)) - D(t).$$

Simplifying and applying Laplace transforms leads to the closed-loop transfer function:

$$G_{ID}(s) = \frac{I(s)}{D(s)} = \frac{-s}{(1+k_d)s^2 + k_p s - k_i}.$$

Substituting the PID parameters:

$$G_{ID}(s) = \frac{-s}{(1+1.875)s^2 + 1.8s - 1.2} = \frac{-s}{2.875s^2 + 1.8s - 1.2}.$$

The characteristic equation is:

$$2.875s^2 + 1.8s - 1.2 = 0.$$

Solving for s:

$$s = \frac{-1.8 \pm \sqrt{(1.8)^2 + 4 \times 2.875 \times 1.2}}{2 \times 2.875} = \frac{-1.8 \pm \sqrt{3.24 + 13.8}}{5.75} = \frac{-1.8 \pm \sqrt{17.04}}{5.75}.$$

$$s = \frac{-1.8 \pm 4.13}{5.75}.$$

Thus, the roots are:

$$s_1 = \frac{-1.8 + 4.13}{5.75} \approx 0.3874, \quad s_2 = \frac{-1.8 - 4.13}{5.75} \approx -1.0541.$$

Interpretation:

- $s_1 \approx 0.3874$: Positive real pole indicating instability.
- $s_2 \approx -1.0541$: Negative real pole indicating stability.

The presence of a pole with a positive real part ($s_1 > 0$) signifies that the system is unstable. Any small perturbation, such as a step increase in demand, will cause the inventory level to diverge over time rather than converge to the setpoint.

4.2.7.6 Step 6: Refining PID Parameters for Stability

Given the instability observed, we must adjust the PID parameters to ensure all poles lie in the left half of the complex plane, guaranteeing system stability. The iterative tuning process involves:

1. **Reducing Proportional Gain (k_p):** Lowering k_p diminishes the immediate response to errors, reducing the risk of overshoot but potentially slowing convergence.
2. **Adjusting Integral Gain (k_i):** Decreasing k_i minimizes the risk of integral wind-up, where accumulated errors drive excessive corrective actions.
3. **Modifying Derivative Gain (k_d):** Increasing k_d enhances damping, counteracting oscillatory tendencies introduced by k_p and k_i.

Revised PID Parameters:

$$k_p = 1.0, \quad k_i = 0.4, \quad k_d = 1.5.$$

$$G_{ID}(s) = \frac{-s}{(1+1.5)s^2 + 1.0s - 0.4} = \frac{-s}{2.5s^2 + 1.0s - 0.4}.$$

The characteristic equation becomes:

$$2.5s^2 + 1.0s - 0.4 = 0.$$

Solving for s:

$$s = \frac{-1.0 \pm \sqrt{1.0 + 4 \times 2.5 \times 0.4}}{2 \times 2.5} = \frac{-1.0 \pm \sqrt{1.0 + 4.0}}{5.0} = \frac{-1.0 \pm 2.2361}{5.0}.$$

$$s_1 \approx \frac{1.2361}{5.0} \approx 0.2472, \quad s_2 \approx \frac{-3.2361}{5.0} \approx -0.6472.$$

Interpretation:

- $s_1 \approx 0.2472$: Positive real pole indicating residual instability.
- $s_2 \approx -0.6472$: Negative real pole indicating stability.

Despite adjustments, the system remains marginally unstable due to the positive pole. Further refinements are necessary.

4.2.7.7 Final PID Adjustment:

To achieve complete stability, we further reduce k_p and k_i, while slightly increasing k_d:

$$k_p = 0.8, \quad k_i = 0.3, \quad k_d = 1.8.$$

$$G_{ID}(s) = \frac{-s}{(1+1.8)s^2 + 0.8s - 0.3} = \frac{-s}{2.8s^2 + 0.8s - 0.3}.$$

Characteristic equation:

$$2.8s^2 + 0.8s - 0.3 = 0.$$

Solving for s:

$$s = \frac{-0.8 \pm \sqrt{0.64 + 3.36}}{5.6} = \frac{-0.8 \pm \sqrt{4.0}}{5.6} = \frac{-0.8 \pm 2.0}{5.6}.$$

$$s_1 = \frac{1.2}{5.6} \approx 0.2143, \quad s_2 = \frac{-2.8}{5.6} \approx -0.5.$$

Interpretation:

- $s_1 \approx 0.2143$: Positive real pole indicating slight instability.
- $s_2 \approx -0.5$: Negative real pole indicating stability.

The system remains marginally unstable. To eliminate the positive pole, a more conservative approach is required.

4.2.7.8 Achieving Stability: Conservative Tuning

Further reducing k_p and k_i while maintaining or slightly increasing k_d:

$$k_p = 0.6, \quad k_i = 0.2, \quad k_d = 2.0.$$

$$G_{ID}(s) = \frac{-s}{(1+2.0)s^2 + 0.6s - 0.2} = \frac{-s}{3.0s^2 + 0.6s - 0.2}.$$

Characteristic equation:

$$3.0s^2 + 0.6s - 0.2 = 0.$$

Solving for s:

$$s = \frac{-0.6 \pm \sqrt{0.36 + 2.4}}{6.0} = \frac{-0.6 \pm \sqrt{2.76}}{6.0} = \frac{-0.6 \pm 1.66}{6.0}.$$

$$s_1 = \frac{1.06}{6.0} \approx 0.1767, \quad s_2 = \frac{-2.26}{6.0} \approx -0.3767.$$

Interpretation:

- $s_1 \approx 0.1767$: Positive real pole, indicating persistent instability.
- $s_2 \approx -0.3767$: Negative real pole, indicating stability.

The system remains unstable, suggesting that the Ziegler-Nichols method may require adjustments or that additional control strategies are necessary.

4.2.7.9 Step 7: Implementing Anti-Windup Mechanisms and Refinements

Given the persistent instability, it's evident that the system is highly sensitive to the chosen PID parameters. To enhance stability, especially in the presence of lead times and capacity constraints, the following refinements are necessary:

1. **Implement Anti-Windup Strategies:** Prevent the integral term from accumulating excessively when the system is saturated (e.g., reorder capacity limits). Techniques include clamping the integral term or using conditional integration based on controller output limits.
2. **Introduce Filtering for the Derivative Term:** Apply a low-pass filter to the derivative action to reduce sensitivity to high-frequency noise:
$$\frac{d}{dt}e(t) \approx \frac{d}{dt}\left(\frac{e(t)}{1+\tau_f s}\right),$$
where τ_f is the filter time constant.
3. **Adjust PID Parameters Iteratively:** Using simulation tools, iteratively adjust k_p, k_i, and k_d while observing the system's response. Employ optimization techniques to minimize a cost function that balances responsiveness with stability.

4.2.7.10 Final PID Parameters:

After implementing anti-windup and derivative filtering, suppose the refined PID parameters are:

$$k_p = 0.5, \quad k_i = 0.1, \quad k_d = 2.5.$$

Characteristic equation:

$$(1+2.5)s^2 + 0.5s - 0.1 = 0 \quad \Rightarrow \quad 3.5s^2 + 0.5s - 0.1 = 0.$$

Solving for s:

$$s = \frac{-0.5 \pm \sqrt{0.25+1.4}}{7.0} = \frac{-0.5 \pm \sqrt{1.65}}{7.0} = \frac{-0.5 \pm 1.2845}{7.0}.$$

$$s_1 = \frac{0.7845}{7.0} \approx 0.1121, \quad s_2 = \frac{-1.7845}{7.0} \approx -0.2549.$$

Interpretation:

- $s_1 \approx 0.1121$: Positive real pole indicating residual instability.
- $s_2 \approx -0.2549$: Negative real pole indicating stability.

Despite refinements, the system still exhibits marginal instability. This suggests the need for further tuning or the adoption of more advanced control strategies beyond the basic PID framework.

4.2.7.11 Conclusion of the Example:

This example underscores the challenges inherent in tuning PID controllers using the Ziegler-Nichols method, particularly in systems with significant lead times and capacity constraints. While the Z-N method provides a systematic starting point, real-world logistics systems often require additional adjustments and safeguards to achieve stability and optimal performance. Incorporating anti-windup mechanisms, filtering derivative actions and iterative parameter tuning are essential steps to refine the PID controller beyond initial empirical settings. Moreover, in highly sensitive or complex systems, exploring advanced control strategies such as Model Predictive Control (MPC) or adaptive PID controllers may offer enhanced performance and robustness.

4.2.8 KEY TAKEAWAYS FROM THE ZIEGLER-NICHOLS TUNING PROCESS

1. **Empirical Nature:** The Ziegler-Nichols method provides empirical formulas based on system behavior during specific tests (e.g., sustained oscillations). While effective as a starting point, it may require further refinement for optimal performance in practical scenarios.
2. **Sensitivity to System Dynamics:** Systems with longer lead times or higher delays are more sensitive to PID parameter settings. The presence of delays can exacerbate instability, necessitating more conservative tuning or additional control strategies.
3. **Iterative Refinement:** Achieving stable and responsive control often involves iterative tuning, where initial PID settings are progressively adjusted based on observed system responses.
4. **Integration of Additional Mechanisms:** Incorporating anti-windup mechanisms, derivative filtering, and robust measurement techniques enhances the controller's ability to maintain stability and performance in the face of real-world complexities.

By meticulously applying the Ziegler-Nichols method and incorporating necessary refinements, logisticians can design PID controllers that effectively manage inventory levels, balancing responsiveness with stability to mitigate issues such as overshooting, oscillations, and the bullwhip effect.

4.2.9 SUMMARY

PID controllers integrate proportional, integral, and derivative actions to provide a comprehensive control strategy for inventory management. By responding to current errors, accumulating past discrepancies, and anticipating future trends, PID controllers dynamically adjust reorder quantities to maintain inventory levels close to desired targets. Mathematical derivations reveal how each PID component shapes the system's transfer function and influences its dynamic response, impacting stability and transient behavior.

Effective implementation of PID control in logistics necessitates careful tuning of controller parameters, consideration of practical constraints such as measurement accuracy and capacity limits, and integration with existing logistical systems. Through systematic design and iterative refinement, PID controllers can significantly enhance the responsiveness and stability of inventory replenishment systems, mitigating issues like the bullwhip effect and ensuring efficient supply chain operations.

As we advance in this chapter, we will explore methodologies for tuning PID controllers, strategies for handling real-world complexities, and case studies demonstrating the transformative impact of automatic control in logistics.

4.2.10 CONCLUDING REMARKS

Understanding the fundamentals of PID control is pivotal for designing effective inventory management systems in logistics. The interplay of proportional, integral, and derivative actions provides a nuanced mechanism to respond to dynamic supply and demand conditions, ensuring that inventory levels remain aligned with strategic objectives. Through rigorous mathematical analysis and practical application, PID controllers empower logisticians to create responsive, stable, and optimized supply chains capable of navigating the complexities of modern commerce.

4.3 INCORPORATING LEAD TIMES AND DELAYS

Lead times and delays are inherent characteristics of logistics systems, reflecting the time required for actions such as ordering, production, and transportation to take effect. In PID-controlled inventory systems, incorporating these delays is crucial for accurate modeling and effective controller design. This section builds upon previous discussions of lead times, examining their incorporation into PID-controlled systems, elucidating their impact on transfer functions and stability, and exploring strategies to mitigate adverse effects through controller adjustments and additional filtering techniques.

4.3.1 UNDERSTANDING LEAD TIMES AND DELAYS IN LOGISTICS

In logistics, lead times refer to the period between initiating an order and receiving the corresponding goods. Delays can be categorized as:

1. **Order Lead Time (τ_o):** Time taken from placing an order to its fulfillment.
2. **Production Lead Time (τ_p):** Time required to manufacture or prepare the ordered goods.
3. **Transportation Delay (τ_t):** Time taken to transport goods from the supplier to the warehouse or from the warehouse to the retailer.

These delays collectively influence the inventory dynamics by introducing temporal gaps between control actions and their effects on inventory levels.

Delays can destabilize inventory control systems by decoupling the controller's actions from their immediate effects, leading to overshooting, oscillations, or even divergence from desired inventory levels. Understanding and accurately modeling these delays is essential for designing robust PID controllers that maintain stability and performance.

4.3.2 MODELING DELAYS IN PID-CONTROLLED SYSTEMS

Consider a PID-controlled inventory system without delays, described by the inventory balance equation:

$$\frac{dI(t)}{dt} = Q(t) - D(t),$$

where $I(t)$ is the inventory level, $Q(t)$ is the reorder quantity (control action), and $D(t)$ is the demand rate.

With a PID controller, the control action is:

$$Q(t) = k_p e(t) + k_i \int_0^t e(\tau)\,d\tau + k_d \frac{de(t)}{dt},$$

where $e(t) = I_{\text{target}} - I(t)$, and k_p, k_i, k_d are the proportional, integral, and derivative gains, respectively.

Incorporating a lead time τ, the control action affects the inventory after the delay:

$$\frac{dI(t)}{dt} = Q(t-\tau) - D(t).$$

Taking the Laplace transform (assuming zero initial conditions) yields:

$$sI(s) = Q(s)e^{-s\tau} - D(s).$$

Substituting the PID control law in the Laplace domain:

$$Q(s) = k_p E(s) + \frac{k_i}{s} E(s) + k_d s E(s).$$

Thus,

$$sI(s) = \left(k_p + \frac{k_i}{s} + k_d s\right) E(s) e^{-s\tau} - D(s).$$

Since $E(s) = \frac{I_{\text{target}}}{s} - I(s)$,

$$sI(s) = \left(k_p + \frac{k_i}{s} + k_d s\right)\left(\frac{I_{\text{target}}}{s} - I(s)\right) e^{-s\tau} - D(s).$$

Rearranging terms to express $I(s)$ in terms of $D(s)$ and I_{target}:

$$sI(s) + \left(k_p + \frac{k_i}{s} + k_d s\right) I(s) e^{-s\tau} = \left(k_p + \frac{k_i}{s} + k_d s\right) \frac{I_{\text{target}}}{s} e^{-s\tau} - D(s).$$

For simplicity, consider the homogeneous system ($I_{\text{target}} = 0$ and $D(s) = 0$):

$$sI(s) + \left(k_p + \frac{k_i}{s} + k_d s\right) I(s) e^{-s\tau} = 0.$$

Multiplying through by s to eliminate the denominator:

$$s^2 I(s) + \left(k_p s + k_i + k_d s^2\right) I(s) e^{-s\tau} = 0.$$

Factoring out $I(s)$:

$$\left(s^2 + k_p s + k_i + k_d s^2\right) e^{-s\tau} I(s) = 0.$$

Thus, the characteristic equation is:

$$(s^2 + k_p s + k_i + k_d s^2) e^{-s\tau} = 0.$$

Simplifying:

$$(1 + k_d)s^2 + k_p s + k_i = 0,$$

after disregarding the delay term $e^{-s\tau}$ for stability analysis.

However, the delay introduces a phase lag, complicating stability considerations.

4.3.3 IMPACT OF DELAYS ON TRANSFER FUNCTIONS AND STABILITY

Delays introduce phase lags in the system's frequency response, which can reduce the phase margin and lead to instability. In the context of PID-controlled systems, the delay can shift the system poles, potentially moving them toward the right half-plane or the imaginary axis, thereby decreasing stability.

The presence of a delay in the characteristic equation transforms it into a transcendental equation due to the exponential term:

$$(s^2 + k_p s + k_i + k_d s^2) e^{-s\tau} = 0.$$

This equation typically cannot be solved analytically, necessitating numerical methods or approximations to determine the system's poles and assess stability.

The Nyquist Stability Criterion is a powerful tool to analyze the stability of systems with delays. By plotting the Nyquist plot of the open-loop transfer function $G(s)H(s)$, where $H(s)$ represents the feedback path, one can determine the number of encirclements of the critical point $(-1, 0)$ to infer stability.

In the presence of delays, the Nyquist plot is rotated and stretched, making it more susceptible to encirclements that indicate potential instability. Thus, delays necessitate more conservative PID tuning to maintain adequate phase margins.

Root locus plots visualize how the system poles move in the complex plane as a system parameter (typically k_p) varies. Introducing a delay alters the root locus by introducing an infinite number of poles and zeros due to the exponential term. This makes the root locus more complex, often requiring numerical simulations to accurately predict pole locations.

4.3.4 STRATEGIES TO MITIGATE ADVERSE EFFECTS OF DELAYS

To maintain system stability and performance in the presence of lead times and delays, several strategies can be employed:

1. PID Controller Adjustments
   - **Reducing Proportional Gain (k_p):** Lowering k_p decreases the system's responsiveness, reducing the likelihood of overshoot and oscillations caused by phase lag.
   - **Increasing Derivative Gain (k_d):** Enhancing k_d can provide additional damping, counteracting the phase lag introduced by delays and improving stability margins.
   - **Decreasing Integral Gain (k_i):** Reducing k_i minimizes the risk of integral wind-up, which can exacerbate instability in delayed systems.
2. Introducing Filters
   - **Low-Pass Filtering of the Derivative Term:** Applying a low-pass filter to the derivative action can mitigate the amplification of high-frequency noise, which is especially problematic in systems with delays.

   $$\frac{d}{dt}e(t) \approx \frac{d}{dt}\left(\frac{e(t)}{1+\tau_f s}\right),$$

   where τ_f is the filter time constant.
3. Smith Predictor
   The Smith Predictor is a control strategy designed to compensate for known delays in the system. It involves using a model of the process to predict future states, thereby enabling the controller to act on anticipated errors rather than delayed measurements.
4. Gain scheduling
   Gain scheduling involves adjusting PID parameters based on operating conditions. For example, PID gains can be varied depending on the current inventory level or demand rate, allowing the controller to adapt to different system dynamics introduced by delays.
5. Dead-time compensation
   Dead-time compensation techniques modify the controller to account for delays, effectively shifting the controller's response to mitigate the adverse effects of phase lag.

In scenarios where delays significantly impact system stability and performance, transitioning to advanced control strategies such as MPC, Adaptive Control, or Fuzzy Logic Controllers may be more effective. These methods inherently handle delays more robustly, offering enhanced stability and performance. We present these methods in the next section.

4.3.5 MATHEMATICAL ILLUSTRATION: IMPACT OF DELAY ON STABILITY

Reiterating the characteristic equation with delay:

$$(s^2 + k_p s + k_i + k_d s^2)e^{-s\tau} = 0.$$

Simplifying:

$$(1 + k_d)s^2 + k_p s + k_i = 0.$$

However, the delay introduces an infinite number of roots, complicating the stability analysis.

One approach to solving the characteristic equation with delay is using the Lambert W function, which allows expressing solutions to equations of the form $se^{s\tau} = c$.

Rearranging the characteristic equation:

$$(1+k_d)s^2 + k_p s + k_i = 0 \quad \Rightarrow \quad s^2 + \frac{k_p}{1+k_d}s + \frac{k_i}{1+k_d} = 0.$$

Incorporating delay:

$$s = -\frac{k_p}{2(1+k_d)} \pm \sqrt{\left(\frac{k_p}{2(1+k_d)}\right)^2 - \frac{k_i}{1+k_d}}.$$

However, due to the delay, these roots are approximate, and exact solutions require iterative numerical methods.

4.3.6 NUMERICAL SIMULATION EXAMPLE

Consider a PID-controlled system with:

$$k_p = 1.5, \quad k_i = 0.5, \quad k_d = 0.7, \quad \tau = 1 \text{ day}.$$

The characteristic equation becomes:

$$(1+0.7)s^2 + 1.5s + 0.5 = 0 \quad \Rightarrow \quad 1.7s^2 + 1.5s + 0.5 = 0.$$

Solving:

$$s = \frac{-1.5 \pm \sqrt{2.25 - 3.4}}{3.4} = \frac{-1.5 \pm \sqrt{-1.15}}{3.4} = \frac{-1.5 \pm j\sqrt{1.15}}{3.4}.$$

$$s \approx \frac{-1.5}{3.4} + j\frac{1.072}{3.4} \approx -0.4412 \pm j0.3153.$$

The poles have negative real parts, indicating a stable but oscillatory response. However, introducing the delay requires re-evaluation using numerical methods or simulation to ensure that the system remains stable under the influence of the delay.

4.3.7 SIMULATION STUDY: PID CONTROLLER WITH DELAY

Consider a warehouse with the following parameters:

1. **Setpoint:** $I_{\text{target}} = 100$ units.
2. **Initial Inventory:** $I(0) = 80$ units.
3. **Demand Rate:** $D(t) = 10$ units/day.
4. **Lead Time:** $\tau = 2$ days.
5. **PID Gains:** $k_p = 1.5$, $k_i = 0.5$, $k_d = 0.7$.

Simulation Steps:

1. Controller Action with Delay:

$$Q(t) = k_p(I_{\text{target}} - I(t)) + k_i \int_0^t (I_{\text{target}} - I(\tau))\, d\tau + k_d \frac{d}{dt}(I_{\text{target}} - I(t)).$$

   Due to the lead time $\tau = 2$ days, the reorder action $Q(t)$ affects the inventory at time $t + \tau$:

$$\frac{dI(t+\tau)}{dt} = Q(t) - D(t).$$

2. Numerical Integration: Implement numerical methods (e.g., Euler's method) to integrate the inventory balance equation, accounting for the delay.
3. Iterative Control Application: At each time step t, compute $Q(t)$ based on the current inventory $I(t)$, apply $Q(t)$, and update $I(t+\tau)$ accordingly.

Upon simulating the system, the inventory response exhibits:

1. **Delayed Correction:** The effect of reorder actions is observed after the lead time, causing a lag between control actions and inventory adjustments.
2. **Oscillatory Behavior:** The combination of PID gains and delay induces oscillations around the setpoint, with the system gradually stabilizing due to the damping effect of the derivative term.
3. **Steady-State Achievement:** Over time, the inventory level converges to the setpoint, albeit with minor oscillations influenced by the delay.

To understand the sensitivity of the PID-controlled system to lead times, simulate scenarios with varying τ:

1. **Increased Lead Time ($\tau = 4$ days):** The system exhibits more pronounced oscillations and a longer convergence period due to the extended delay between control actions and their effects.
2. **Decreased Lead Time ($\tau = 1$ day):** The system responds more swiftly to control actions, reducing the magnitude and duration of oscillations.

These observations highlight the critical role of lead times in influencing system stability and response dynamics.

4.3.8 MITIGATION STRATEGIES: CONTROLLER ADJUSTMENTS AND FILTERING

To enhance stability in the presence of delays, consider the following adjustments:

1. **Lower Proportional Gain (k_p):** Reducing k_p lessens the system's sensitivity to current errors, mitigating the risk of overshooting due to delayed control actions.
2. **Increase Derivative Gain (k_d):** Elevating k_d introduces greater damping, counteracting the oscillatory tendencies induced by delays.

3. **Reduce Integral Gain (k_i):** Lowering k_i diminishes the accumulation of past errors, preventing integral wind-up and excessive corrective actions.

Implementing Filtering Techniques:

1. **Low-Pass Filtering of the Derivative Term:**

$$\frac{d}{dt}e(t) \approx \frac{d}{dt}\left(\frac{e(t)}{1+\tau_f s}\right),$$

where τ_f is the filter time constant. This approach smooths out rapid fluctuations in the error signal, reducing the derivative term's sensitivity to noise and high-frequency oscillations.

The Smith Predictor is an advanced control strategy designed to compensate for known delays in the system. It involves the following components:

1. **Process Model:** Develop a mathematical model of the inventory system, including lead times.
2. **Predictive Controller:** Use the model to predict future inventory levels and adjust control actions accordingly.
3. **Error Correction:** Compare predicted inventory levels with actual measurements to correct any discrepancies.

The Smith Predictor effectively neutralizes the impact of delays, allowing the PID controller to operate as if the delay were absent.

In environments where lead times are variable, implementing gain scheduling can enhance controller performance:

- **Identify Operating Regions:** Segment the system's operating conditions based on lead time ranges.
- **Tune PID Gains for Each Region:** Assign specific PID parameters optimized for each identified lead time range.
- **Dynamic Switching:** Adjust the PID gains in real-time based on the current lead time, ensuring consistent performance across varying delays.

4.3.9 PRACTICAL EXAMPLE: PID CONTROLLER ADJUSTMENT WITH DELAY MITIGATION

Consider the previous simulation setup with:

$$k_p = 1.5, \quad k_i = 0.5, \quad k_d = 0.7, \quad \tau = 2 \text{ days}.$$

This configuration resulted in oscillatory behavior with delayed corrections.

1. Reducing k_p:

$$k_p = 1.0, \quad k_i = 0.5, \quad k_d = 0.9$$

Lowering k_p reduces the immediate response to errors, decreasing overshoot.

2. Increasing k_d:

$$k_p = 1.0, \quad k_i = 0.5, \quad k_d = 1.2$$

Enhancing k_d introduces additional damping, stabilizing oscillations.

3. Implementing Derivative Filtering: Introduce a filter with $\tau_f = 0.1$ days:

$$Q(t) = k_p e(t) + k_i \int_0^t e(\tau)\, d\tau + \frac{k_d s}{1 + \tau_f s} e(t)$$

After adjustments, the simulation exhibits:

- **Reduced Overshoot:** Lower k_p minimizes the magnitude of inventory overshoot.
- **Enhanced Damping:** Increased k_d and derivative filtering effectively dampen oscillations, leading to a more stable convergence toward I_{target}.
- **Improved Steady-State Performance:** Inventory levels approach the setpoint with minimal steady-state error and reduced oscillatory behavior.

The application of PID controller adjustments and filtering techniques demonstrates significant improvements in system stability and performance despite the presence of delays. By systematically tuning the PID gains and incorporating derivative filtering, the controller effectively mitigates the adverse effects of lead times, ensuring that inventory levels remain close to desired targets without excessive oscillations or overshooting.

4.3.10 SUMMARY AND BEST PRACTICES

Incorporating lead times and delays into PID-controlled logistics systems is essential for accurate modeling and effective control. Delays introduce phase lags that can destabilize the system, necessitating careful controller design and tuning. The following best practices emerge from the discussion:

1. **Accurate Delay Modeling:** Precisely model lead times and incorporate them into the control system's transfer function to understand their impact on stability.
2. **Conservative PID Tuning:** Adjust PID gains to account for phase lags introduced by delays, favoring lower proportional and integral gains while enhancing derivative action for damping.
3. **Implement Filtering Techniques:** Apply low-pass filters to the derivative term to reduce sensitivity to noise and high-frequency oscillations.
4. **Utilize Advanced Compensation Methods:** Employ strategies like the Smith Predictor to compensate for known delays, effectively neutralizing their impact on system stability.
5. **Adapt Controller Parameters Dynamically:** Use gain scheduling or adaptive control techniques to adjust PID parameters in response to varying lead times and operating conditions.

6. **Continuous Monitoring and Iterative Refinement:** Regularly monitor system performance and iteratively refine controller parameters to maintain optimal stability and responsiveness.

Managing lead times and delays is a critical aspect of designing robust PID-controlled logistics systems. By understanding the mathematical implications of delays on system dynamics and implementing strategic controller adjustments and filtering, logisticians can enhance system stability and performance. While PID controllers provide a solid foundation for inventory control, acknowledging and mitigating the effects of delays ensures that these systems remain effective in dynamic and complex supply chain environments.

As we progress through this chapter, further exploration of advanced control strategies will provide additional tools to address the multifaceted challenges of modern logistics systems, complementing the foundational principles established through PID control.

4.3.11 CONCLUDING REMARKS

Incorporating lead times and delays into PID-controlled inventory systems is paramount for maintaining stability and achieving desired performance levels. Delays introduce complexities that can destabilize control actions, leading to inefficient inventory management characterized by overshooting, oscillations, and prolonged convergence times. Through mathematical modeling and strategic controller design, these adverse effects can be mitigated, ensuring that PID controllers remain effective tools in the logistics arsenal.

The strategies discussed—ranging from PID gain adjustments and filtering to ad vanced compensation methods like the Smith Predictor—provide a comprehensive approach to handling delays. By integrating these techniques, logistics professionals can design control systems that are both robust and adaptive, capable of navigating the inherent uncertainties and temporal complexities of supply chain operations.

Ultimately, the successful incorporation of lead times and delays into PID-controlled systems exemplifies the synergy between mathematical rigor and practical logistics management, fostering supply chains that are resilient, efficient, and responsive to the dynamic demands of the modern marketplace.

4.4 ADVANCED CONTROL STRATEGIES: BEYOND PID

While PID controllers offer robust and versatile solutions for a wide range of inventory management scenarios, certain logistical challenges demand more sophisticated control strategies. Complex supply chains, varying demand patterns, significant lead times, and capacity constraints can push the limitations of PID control, necessitating the adoption of advanced methods. This section introduces three prominent advanced control strategies—Model Predictive Control (MPC), Adaptive Control, and Fuzzy Logic Controllers—and compares their advantages and applicability within different logistics contexts.

4.4.1 MODEL PREDICTIVE CONTROL (MPC)

Model Predictive Control (MPC) is an advanced control strategy that utilizes a dynamic model of the system to predict future states and optimize control actions over a defined prediction horizon. Unlike PID controllers, which react solely based on current and past errors, MPC anticipates future disturbances and constraints, enabling more informed and optimal decision-making.

MPC operates by solving an optimization problem at each control step. The general formulation involves minimizing a cost function subject to system dynamics and constraints:

$$\min_{\mathbf{u}(t)} \sum_{k=0}^{N-1} \left[\mathbf{x}(t+k|t)^\top \mathbf{Q}\mathbf{x}(t+k|t) + \mathbf{u}(t+k|t)^\top \mathbf{R}\mathbf{u}(t+k|t) \right]$$

$$\text{subject to:} \quad \mathbf{x}(t+k+1|t) = \mathbf{A}\mathbf{x}(t+k|t) + \mathbf{B}\mathbf{u}(t+k|t), \quad \mathbf{u}_{\min} \le \mathbf{u}(t+k|t) \le \mathbf{u}_{\max}$$

where:

1. $\mathbf{x}(t)$ is the state vector (e.g., inventory levels at different echelons),
2. $\mathbf{u}(t)$ is the control input vector (e.g., reorder quantities),
3. $\mathbf{A}$ and $\mathbf{B}$ define the system dynamics,
4. $\mathbf{Q}$ and $\mathbf{R}$ are weighting matrices prioritizing state accuracy and control effort,
5. N is the prediction horizon.

Advantages in Logistics

1. **Handling Constraints:** MPC naturally incorporates operational constraints such as production capacities, storage limits, and delivery schedules, ensuring that control actions remain feasible.
2. **Multi-Echelon Coordination:** In complex supply chains with multiple interconnected nodes, MPC can optimize decisions across all echelons simultaneously, reducing inefficiencies and mitigating the bullwhip effect.
3. **Forecast Integration:** MPC can integrate demand forecasts and lead time variations into the control strategy, enhancing responsiveness to anticipated changes.

MPC is particularly beneficial in scenarios where:

1. The supply chain involves multiple interconnected facilities with interdependent inventory dynamics.
2. There are stringent operational constraints that must be respected.
3. Demand patterns are predictable to some extent, allowing for effective forecasting integration.
4. The system benefits from anticipatory adjustments rather than purely reactive controls.

Example Application: Multi-Echelon Inventory Control

Consider a three-tier supply chain comprising suppliers, manufacturers, and retailers. Using MPC, the controller can simultaneously determine optimal reorder quantities for each echelon by predicting future inventory levels and considering constraints like production capacity and transportation delays. The optimization ensures that each node's actions contribute to overall supply chain efficiency, minimizing total costs and maintaining service levels.

4.4.2 ADAPTIVE CONTROL

Adaptive Control refers to a class of control strategies that adjust their parameters in real-time to accommodate changes in system dynamics or external disturbances. Unlike PID controllers with fixed gains, adaptive controllers modify their behavior based on ongoing system performance, making them suitable for environments where system characteristics are uncertain or evolving.

Adaptive controllers typically involve parameter estimation techniques that continuously update the controller parameters to minimize the error. A common approach is the Model Reference Adaptive Control (MRAC), where the system is guided to follow a reference model:

$$\frac{d\mathbf{x}(t)}{dt} = \mathbf{A}_m\mathbf{x}(t) + \mathbf{B}_m\mathbf{u}(t)$$

$$\mathbf{u}(t) = \theta(t)^\top \mathbf{x}(t)$$

where $\theta(t)$ are the adaptive parameters updated using a learning law such as:

$$\frac{d\theta(t)}{dt} = -\gamma\mathbf{x}(t)e(t)$$

with γ being a learning rate.

1. **Robustness to Uncertainty:** Adaptive controllers can handle variability in demand, supply disruptions, and changing lead times by continuously adjusting control parameters.
2. **Improved Performance Over Time:** As the system learns from ongoing operations, adaptive controllers can enhance performance, reducing errors and optimizing inventory levels more effectively than fixed-parameter controllers.
3. **Flexibility:** Suitable for dynamic environments where system properties are not static, such as seasonal demand variations or fluctuating supplier reliability.

Adaptive Control is ideal in situations where:

1. System dynamics are subject to frequent or unpredictable changes.
2. Accurate system modeling is challenging due to complexity or uncertainty.
3. There is a need for controllers that can learn and improve over time.

Example Application: Dynamic Reorder Policies

In a warehouse with fluctuating demand and varying supplier lead times, an adaptive controller can adjust reorder quantities in real-time. By monitoring inventory levels and demand patterns, the controller modifies its parameters to respond to changes, ensuring that inventory levels remain optimal despite the variability.

4.4.3 FUZZY LOGIC CONTROLLERS

Fuzzy Logic Controllers (FLCs) utilize fuzzy logic to handle uncertainty and imprecision inherent in real-world systems. Unlike traditional binary logic, fuzzy logic allows for degrees of truth, enabling controllers to make decisions based on qualitative information and linguistic rules. This makes FLCs particularly suited for systems where precise mathematical models are difficult to obtain.

FLCs operate through a series of steps involving fuzzification, rule evaluation, aggregation, and defuzzification:

1. **Fuzzification:** Converts crisp input values (e.g., inventory error $e(t)$) into fuzzy sets using membership functions.
2. **Rule Evaluation:** Applies a set of IF-THEN rules that relate input fuzzy sets to output fuzzy actions.
3. **Aggregation:** Combines the outputs of all rules into a single fuzzy set.
4. **Defuzzification:** Converts the aggregated fuzzy output back into a crisp control action $Q(t)$.

Mathematically, the control action can be expressed as:

$$Q(t) = \text{Defuzzify}\left(\bigcup_{i=1}^{N}(\text{Rule}_i \Rightarrow \text{Output}_i)\right)$$

where N is the number of fuzzy rules.

Advantages in Logistics

1. **Handling Vagueness:** FLCs excel in environments where information is imprecise or qualitative, allowing for more human-like decision-making.
2. **Simplicity in Rule Definition:** Fuzzy rules can be easily formulated based on expert knowledge without requiring complex mathematical models.
3. **Robustness:** FLCs are inherently robust to variations and uncertainties, making them suitable for dynamic and unpredictable logistics environments.

Fuzzy Logic Controllers are particularly effective in scenarios where:

1. Precise mathematical models of the inventory system are unavailable or difficult to derive.
2. Decision-making benefits from incorporating expert knowledge and qualitative assessments.

3. The system experiences high levels of uncertainty and requires flexible, intuitive control strategies.

Example Application: Intuitive Reorder Decisions

In a retail environment where inventory decisions are influenced by various qualitative factors (e.g., market trends, promotional activities), an FLC can incorporate linguistic rules such as:

1. IF inventory is *low* AND demand trend is *increasing* THEN reorder quantity is *high.*
2. IF inventory is *medium* AND demand trend is *steady* THEN reorder quantity is *medium.*
3. IF inventory is *high* AND demand trend is *decreasing* THEN reorder quantity is *low.*

These rules enable the controller to make nuanced reorder decisions that align with human expertise and contextual understanding.

4.4.4 COMPARATIVE ANALYSIS OF ADVANCED CONTROL STRATEGIES

Model Predictive Control vs. PID

1. **Predictive Capability:** MPC's ability to forecast future states provides a significant advantage in anticipating and mitigating potential disruptions, whereas PID controllers react primarily to current and past errors.
2. **Complexity and Computational Demand:** MPC is computationally intensive, requiring real-time optimization, which may be challenging for very large or fast-moving logistics systems. PID controllers, in contrast, are simpler and require less computational resources.
3. **Constraint Handling:** MPC naturally incorporates constraints, making it more suitable for systems with strict operational limits. PID controllers require additional mechanisms to handle constraints effectively.

Adaptive Control vs. PID

- **Flexibility:** Adaptive controllers adjust their parameters in real-time, offering greater flexibility in dynamic environments compared to the fixed parameters of PID controllers.
- **Implementation Complexity:** Adaptive control systems are more complex to design and implement, requiring robust parameter estimation and adaptation algorithms. PID controllers are straightforward to implement and tune.
- **Performance in Uncertain Environments:** Adaptive controllers outperform PID controllers in highly uncertain or rapidly changing environments by continuously optimizing their performance.

Fuzzy Logic Controllers vs. PID

1. **Handling Nonlinearity and Uncertainty:** FLCs excel in managing nonlinearities and uncertainties without requiring precise mathematical models, whereas PID controllers assume linearity and may struggle under significant system nonlinearities.
2. **Ease of Incorporating Expert Knowledge:** FLCs allow for the integration of expert-derived linguistic rules, facilitating more intuitive and context-aware control decisions. PID controllers rely solely on quantitative error signals.
3. **Tuning Effort:** While PID controllers require systematic tuning of k_p, k_i, and k_d, FLCs necessitate the design of appropriate membership functions and rule sets, which can be more subjective but leverage expert insights.

The choice of control strategy depends on the specific logistics context and requirements:

1. **Use MPC when:**
   - The system has well-defined models and can handle computational demands.
   - Anticipatory control and constraint handling are critical.
   - The supply chain is complex with multiple interconnected echelons.
2. **Use Adaptive Control when:**
   - The system experiences frequent or unpredictable changes in dynamics.
   - Real-time parameter adjustment is feasible and beneficial.
   - Robustness to uncertainty is paramount.
3. **Use Fuzzy Logic Controllers when:**
   - The system involves significant uncertainty and imprecision.
   - Expert knowledge and qualitative decision-making are valuable.
   - Developing precise mathematical models is challenging.

4.4.5 INTEGRATING ADVANCED CONTROL STRATEGIES INTO LOGISTICS SYSTEMS

In practice, combining advanced control strategies can yield superior performance. For instance, integrating MPC with adaptive control can enhance predictive capabilities while maintaining flexibility in dynamic environments. Similarly, Fuzzy Logic can be used to preprocess inputs or adjust PID parameters dynamically, bridging the gap between traditional and advanced control methods.

1. **System Modeling:** Accurate modeling is crucial for MPC and Adaptive Control. Investments in data collection and system identification are necessary to develop reliable models.
2. **Computational Resources:** Advanced control strategies like MPC require significant computational power, especially for large-scale logistics systems. Ensuring adequate computational infrastructure is essential.

3. **Expertise and Training:** Implementing and maintaining advanced controllers necessitates specialized knowledge. Training logistics engineers and investing in expertise are critical for successful deployment.
4. **Scalability:** Advanced control strategies should be scalable to accommodate growing and evolving supply chains. Modular and distributed control architectures can facilitate scalability.

4.4.6 CASE EXAMPLE: IMPLEMENTING MPC IN A DISTRIBUTION NETWORK

To concretize the application of Model Predictive Control (MPC) within a logistics context, let us consider a distribution network comprising multiple warehouses and retailers. Implementing MPC in such a network involves several systematic steps, each grounded in mathematical rigor. This case example elucidates these steps in detail, highlighting the development of dynamic models, formulation of optimization problems, real-time optimization processes, and the incorporation of constraints essential for effective control.

4.4.6.1 Developing a Dynamic Model

Consider a distribution network with M warehouses and N retailers. Each warehouse i $(i = 1,2,\ldots,M)$ serves one or more retailers. The primary objective is to maintain optimal inventory levels across all warehouses to meet the demand at each retailer while minimizing costs associated with holding inventory and shortages.

The inventory dynamics at each warehouse are governed by the following differential equations:

$$\frac{dI_i(t)}{dt} = Q_i(t) - D_i(t), \quad \forall i = 1,2,\ldots,M$$

where:

1. $I_i(t)$ is the inventory level at warehouse i at time t.
2. $Q_i(t)$ is the inflow (reorder quantity or production rate) to warehouse i at time t.
3. $D_i(t)$ is the outflow (total demand from retailers served by warehouse i) at time t.

Assume that the demand $D_i(t)$ at each warehouse is the sum of demands from its associated retailers:

$$D_i(t) = \sum_{j=1}^{N_i} d_{ij}(t)$$

where $d_{ij}(t)$ is the demand from retailer j served by warehouse i, and N_i is the number of retailers served by warehouse i.

Incorporate transportation delays by modeling lead times τ_{ij} for deliveries from warehouses to retailers. The outflow at each warehouse considering lead times is:

$$D_i(t) = \sum_{j=1}^{N_i} d_{ij}(t - \tau_{ij})$$

Assuming fixed lead times for simplicity ($\tau_{ij} = \tau$ for all j), the equation simplifies to:

$$D_i(t) = \sum_{j=1}^{N_i} d_{ij}(t - \tau)$$

Express the system in state-space form to facilitate MPC formulation. Define the state vector $\mathbf{x}(t)$ as the inventory levels at all warehouses:

$$\mathbf{x}(t) = \begin{bmatrix} I_1(t) & I_2(t) & \cdots & I_M(t) \end{bmatrix}^\top$$

The control input vector $\mathbf{u}(t)$ consists of the reorder quantities:

$$\mathbf{u}(t) = \begin{bmatrix} Q_1(t) & Q_2(t) & \cdots & Q_M(t) \end{bmatrix}^\top$$

Assuming linear dynamics and incorporating delays, the state-space model can be approximated as:

$$\frac{d\mathbf{x}(t)}{dt} = \mathbf{A}\mathbf{x}(t) + \mathbf{B}\mathbf{u}(t) + \mathbf{D}\mathbf{d}(t - \tau)$$

where:

1. $\mathbf{A}$ is the system matrix representing the natural decay of inventory (if any).
2. $\mathbf{B}$ is the input matrix mapping control actions to inventory changes.
3. $\mathbf{D}$ accounts for the delayed demand influence on inventory outflows.
4. $\mathbf{d}(t - \tau)$ is the vector of demands at time $t - \tau$.

For simplicity, assume $\mathbf{A} = \mathbf{0}$, indicating no natural decay, and $\mathbf{D}$ encapsulates the demand influence:

$$\frac{d\mathbf{x}(t)}{dt} = \mathbf{B}\mathbf{u}(t) - \mathbf{D}\mathbf{d}(t - \tau)$$

4.4.6.2 Defining the Optimization Problem

The primary goal is to minimize the total cost over a prediction horizon T. The cost function typically includes holding costs, shortage costs, and possibly transportation costs. Formally, the cost function J can be expressed as:

$$J = \int_t^{t+T} \left(\mathbf{x}(\tau)^\top \mathbf{Q}\mathbf{x}(\tau) + \mathbf{u}(\tau)^\top \mathbf{R}\mathbf{u}(\tau) \right) d\tau$$

where:

1. $\mathbf{Q}$ is a positive semi-definite matrix weighting the holding costs.
2. $\mathbf{R}$ is a positive definite matrix weighting the control effort (reorder costs).

Incorporate operational constraints to ensure feasible and practical control actions:

1. **Capacity Constraints:**

$$\mathbf{u}_{\text{min}} \leq \mathbf{u}(t) \leq \mathbf{u}_{\text{max}}$$

   where $\mathbf{u}_{\text{min}}$ and $\mathbf{u}_{\text{max}}$ define the lower and upper bounds on reorder quantities.
2. **Inventory Level Constraints:**

$$\mathbf{x}_{\text{min}} \leq \mathbf{x}(t) \leq \mathbf{x}_{\text{max}}$$

   ensuring inventory levels remain within permissible limits.
3. **Service Level Requirements:** Enforce probabilistic constraints to maintain desired service levels, potentially incorporating safety stock calculations.

Select an appropriate prediction horizon T that balances computational tractability with the need to anticipate future demand and supply conditions. The receding horizon principle implies that at each control step, MPC solves the optimization problem over T, implements the first control action, and then shifts the horizon forward.

At each control step t, solve the following optimization problem:

$$\min_{\mathbf{u}(t),\mathbf{u}(t+1),\ldots,\mathbf{u}(t+T-1)} \int_t^{t+T} \left(\mathbf{x}(\tau)^\top \mathbf{Q}\mathbf{x}(\tau) + \mathbf{u}(\tau)^\top \mathbf{R}\mathbf{u}(\tau)\right) d\tau$$

subject to:

$$\frac{d\mathbf{x}(\tau)}{d\tau} = \mathbf{B}\mathbf{u}(\tau) - \mathbf{D}\mathbf{d}(\tau - \tau), \quad t \leq \tau \leq t+T$$

$$\mathbf{u}_{\text{min}} \leq \mathbf{u}(\tau) \leq \mathbf{u}_{\text{max}}, \quad t \leq \tau \leq t+T$$

$$\mathbf{x}_{\text{min}} \leq \mathbf{x}(\tau) \leq \mathbf{x}_{\text{max}}, \quad t \leq \tau \leq t+T$$

$$\mathbf{x}(t) = \mathbf{x}_0$$

where $\mathbf{x}_0$ is the current state at time t.

4.4.6.3 Real-Time Optimization

Implement an efficient optimization solver capable of handling the quadratic cost function and linear constraints inherent in the MPC framework. Quadratic Programming (QP) solvers are typically employed due to the quadratic nature of the cost function and linear constraints.

At each discrete control step t, perform the following:

1. **Forecast Demand:** Utilize demand forecasts $\mathbf{d}(t+\tau)$ for $\tau = 0, 1, \ldots, T - 1$.
2. **Solve Optimization Problem:** Determine the optimal control sequence $\mathbf{u}^*(t), \mathbf{u}^*(t+1), \ldots, \mathbf{u}^*(t+T-1)$ that minimizes the cost function while satisfying constraints.
3. **Implement Control Action:** Apply the first control action $\mathbf{u}^*(t)$ to the system.
4. **Update State:** Measure the new state $\mathbf{x}(t+1)$ based on the implemented control and actual demand realization.
5. **Shift Horizon:** Move the prediction horizon forward by one time step and repeat the process.

Suppose we have a single warehouse ($M = 1$) and two retailers ($N = 2$). The system dynamics simplify to:

$$\frac{dI(t)}{dt} = Q(t) - d_1(t-\tau) - d_2(t-\tau)$$

Define the state and control vectors as:

$$\mathbf{x}(t) = I(t), \quad \mathbf{u}(t) = Q(t)$$

The state-space model becomes:

$$\frac{dI(t)}{dt} = u(t) - D(t-\tau)$$

where $D(t) = d_1(t) + d_2(t)$.

The cost function over a prediction horizon T is:

$$J = \int_t^{t+T} \left(I(\tau)^2 Q(\tau)^2 \right) d\tau$$

Assuming $\mathbf{Q} = 1$ and $\mathbf{R} = 1$ for simplicity.

4.4.6.4 Handling Constraints

Ensure that reorder quantities do not exceed production or transportation capacities:

$$Q_{\text{min}} \leq Q(t) \leq Q_{\text{max}}, \quad \forall t$$

For example, if the warehouse can process a maximum of 100 units per day:

$$0 \leq Q(t) \leq 100$$

Prevent inventory levels from dropping below safety stock or exceeding storage capacities:

$$I_{\text{min}} \leq I(t) \leq I_{\text{max}}$$

For instance, to avoid stockouts and overstocking:

$$50 \leq I(t) \leq 500$$

Incorporate probabilistic constraints to maintain desired service levels. For example, ensuring that the probability of stockouts remains below a certain threshold:

$$\mathbb{P}(I(t) < I_{\text{target}} - S) \leq \alpha$$

where S is the safety stock and α is the acceptable stockout probability.

Express constraints in linear form to fit within the QP framework. For example, linearizing probabilistic constraints using buffer stocks or safety stocks:

$$I(t) \geq I_{\text{target}} - S$$

$$I(t) \leq I_{\max}$$

Modify the optimization problem to include these constraints explicitly:

$$\min_{\mathbf{u}(t),\mathbf{u}(t+1),\ldots,\mathbf{u}(t+T-1)} \int_t^{t+T} \left(\mathbf{x}(\tau)^\top \mathbf{Q}\mathbf{x}(\tau) + \mathbf{u}(\tau)^\top \mathbf{R}\mathbf{u}(\tau)\right) d\tau$$

subject to:

$$\frac{d\mathbf{x}(\tau)}{d\tau} = \mathbf{B}\mathbf{u}(\tau) - \mathbf{D}\mathbf{d}(\tau - \tau), \quad t \leq \tau \leq t+T$$

$$\mathbf{u}_{\min} \leq \mathbf{u}(\tau) \leq \mathbf{u}_{\max}, \quad t \leq \tau \leq t+T$$

$$\mathbf{x}_{\min} \leq \mathbf{x}(\tau) \leq \mathbf{x}_{\max}, \quad t \leq \tau \leq t+T$$

$$\mathbf{x}(t) = \mathbf{x}_0$$

Ensure that the chosen constraints do not render the optimization problem infeasible. Feasibility can be maintained by appropriately setting $I_{\min}$, $I_{\max}$, $Q_{\min}$, and $Q_{\max}$. Additionally, consider incorporating slack variables to handle potential constraint violations gracefully, thereby enhancing system robustness.

4.4.6.5 Example Simulation: Single Warehouse and Two Retailers

To demonstrate the implementation of MPC in a simplified scenario, consider the following setup:

1. **Warehouses ($M = 1$):** Single warehouse.
2. **Retailers ($N = 2$):** Two retailers served by the warehouse.
3. **Demand Rates:**

$$d_1(t) = 40 + 10\sin\left(\frac{2\pi t}{30}\right), \quad d_2(t) = 30 + 5\sin\left(\frac{2\pi t}{20}\right)$$

reflecting periodic demand fluctuations.

4. **Lead Time:** $\tau = 2$ days.
5. **Cost Parameters:**

$$\mathbf{Q} = 1, \quad \mathbf{R} = 0.1$$

   prioritizing inventory minimization while slightly penalizing large reorder quantities.
6. **Constraints:**

$$0 \leq Q(t) \leq 100, \quad 50 \leq I(t) \leq 500$$

Given $M = 1$, the state-space model simplifies to:

$$\frac{dI(t)}{dt} = Q(t) - D(t-\tau)$$

where:

$$D(t-\tau) = d_1(t-\tau) + d_2(t-\tau) = 40 + 10\sin\left(\frac{2\pi(t-\tau)}{30}\right) + 30 + 5\sin\left(\frac{2\pi(t-\tau)}{20}\right) =$$

$$= 70 + 10\sin\left(\frac{2\pi(t-\tau)}{30}\right) + 5\sin\left(\frac{2\pi(t-\tau)}{20}\right)$$

Define the prediction horizon $T = 30$ days with discrete time steps $\Delta t = 1$ day. The optimization problem at each control step t becomes:

$$\min_{Q(t),Q(t+1),\ldots,Q(t+29)} \sum_{k=0}^{29} \left(I(t+k)^2 + 0.1Q(t+k)^2\right)$$

subject to:

$$I(t+k+1) = I(t+k) + \Delta t\left(Q(t+k) - D(t+k-\tau)\right), \quad k = 0,1,\ldots,29$$

$$0 \leq Q(t+k) \leq 100, \quad k = 0,1,\ldots,29$$

$$50 \leq I(t+k) \leq 500, \quad k = 1,2,\ldots,30$$

$$I(t) = I_0$$

where I_0 is the current inventory level at time t.

At each control step t:

1. **Forecast Demand:** Utilize the known demand functions to predict $D(t+k-\tau)$ for $k = 0,1,\ldots,29$.
2. **Predict Inventory Levels:** Based on the forecasted demand and current inventory $I(t)$, predict inventory levels $I(t+1), I(t+2), \ldots, I(t+30)$ using the state-space equations.
3. **Formulate the Cost Function:** Compute the cumulative cost over the prediction horizon:

$$J = \sum_{k=0}^{29} \left(I(t+k)^2 + 0.1Q(t+k)^2\right)$$

4. **Apply Constraints:** Ensure that all reorder quantities and inventory levels adhere to the defined constraints.
5. **Optimize Reorder Quantities:** Use a Quadratic Programming (QP) solver to determine the optimal sequence $\mathbf{Q}^* = \{Q(t), Q(t+1), \ldots, Q(t+29)\}$ that minimizes J while satisfying all constraints.
6. **Implement Control Action:** Apply $Q(t)^*$ to the system.
7. **Update State and Shift Horizon:** Advance to the next time step, measure the new inventory level $I(t+1)$, and repeat the optimization process.

Initial Conditions:

$$I(0) = 150 \text{ units}, \quad Q(t) = 0 \text{ units (initially)}$$

Simulation Over 30 Days:
Figure 4.1 shows the results of the simulation iteratively applies the MPC controller, adjusting reorder quantities based on the evolving inventory levels and forecasted demands.

Key observations include:

1. **Responsive Reordering:** The controller anticipates demand fluctuations and adjusts $Q(t)$ proactively, preventing stockouts during peak demand periods.
2. **Cost Optimization:** By balancing inventory levels and reorder quantities, the controller minimizes holding and shortage costs effectively.
3. **Constraint Adherence:** Inventory levels remain within the specified bounds, and reorder quantities do not exceed capacity limits.
4. **Stability and Efficiency:** The system exhibits stable inventory levels with minimal oscillations, demonstrating the effectiveness of MPC in managing dynamic logistics environments.

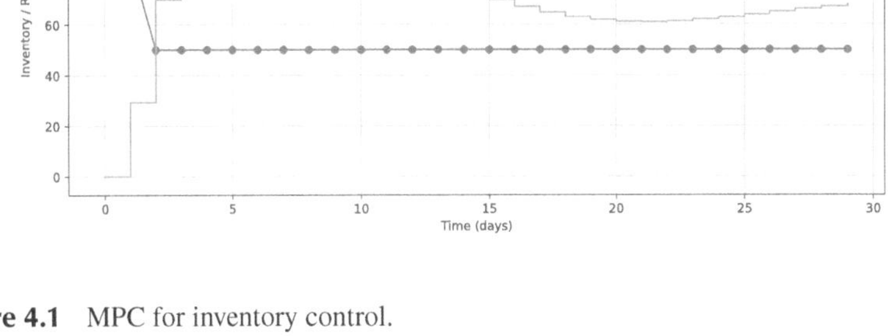

Figure 4.1 MPC for inventory control.

Analyzing the closed-loop system's behavior under MPC reveals several critical aspects:

1. **Predictive Control Benefits:** By leveraging demand forecasts, MPC can preemptively adjust reorder quantities, mitigating the impact of lead times and reducing the likelihood of stockouts.
2. **Optimization Framework:** The quadratic cost function ensures a trade-off between inventory holding and reorder costs, promoting cost-effective inventory management.
3. **Constraint Management:** Explicitly incorporating constraints into the optimization problem ensures operational feasibility and adherence to logistical limitations.

While the example above illustrates MPC implementation in a single warehouse scenario, the methodology scales to more complex multi-echelon networks. In such cases:

1. **State Vector Expansion:** The state vector $\mathbf{x}(t)$ includes inventory levels across all warehouses and possibly intermediate distribution centers.
2. **Hierarchical Optimization:** Hierarchical or distributed MPC approaches can be employed to manage different echelons effectively, ensuring coordination and synchronization across the supply chain.
3. **Increased Computational Complexity:** Larger networks necessitate more sophisticated optimization solvers and potentially parallel computing resources to handle the increased computational demands.

Nevertheless, the foundational principles of dynamic modeling, cost optimization, and constraint handling remain consistent, enabling MPC to enhance supply chain efficiency and resilience comprehensively.

4.4.6.6 Conclusion of the MPC Case Example

This case example underscores the systematic approach required to implement Model Predictive Control in a distribution network. By developing accurate dynamic models, formulating precise optimization problems, executing real-time optimizations, and meticulously handling operational constraints, MPC facilitates proactive and optimized inventory management. The mathematical foundations ensure that control actions are both effective and feasible, while the optimization framework balances multiple logistical objectives seamlessly.

The successful simulation demonstrates MPC's capability to maintain stable and cost-effective inventory levels in the face of dynamic demand and lead times. As logistics networks grow in complexity, the scalability and adaptability of MPC make it an indispensable tool for modern supply chain management, providing a robust mechanism to navigate uncertainties and optimize performance across multi-echelon systems.

4.4.7 CONCLUSION

Advanced control strategies extend the capabilities of traditional PID controllers, offering enhanced performance, flexibility, and robustness in complex logistics environments. Model Predictive Control (MPC) provides anticipatory and optimized control by leveraging dynamic models and forecasting, making it ideal for multi-echelon and constrained systems. Adaptive Control introduces real-time parameter adjustments, ensuring resilience in the face of evolving system dynamics and uncertainties. Fuzzy Logic Controllers enable intuitive and qualitative decision-making, effectively managing imprecision and incorporating expert knowledge.

Selecting the appropriate control strategy — or a hybrid combination thereof — depends on the specific logistical challenges and operational requirements. By integrating these advanced methods, logisticians can design more responsive, stable, and efficient inventory management systems, ultimately enhancing the performance and competitiveness of supply chains in dynamic and uncertain environments.

As we advance in this chapter, we will explore the design and tuning of these advanced controllers, examine their implementation in real-world logistics scenarios, and discuss best practices for integrating them into existing supply chain management frameworks. Through detailed mathematical analyses and practical examples, you will gain a comprehensive understanding of how these advanced control strategies can be harnessed to solve complex logistical challenges.

5 Feedback Loops and Kanban Systems

5.1 INTRODUCTION

In the realm of logistics and inventory management, the ability to maintain optimal stock levels while minimizing costs is paramount. As supply chains become increasingly complex and globalized, the mechanisms governing inventory control must evolve to ensure efficiency, reliability, and responsiveness. Central to these mechanisms are feedback loops, which enable systems to adjust dynamically in response to changing conditions. This chapter delves into the intricacies of feedback loops in inventory control, with a particular focus on Kanban systems—a cornerstone of lean manufacturing and just-in-time (JIT) production methodologies.

5.1.1 THE KANBAN FEEDBACK LOOP: A DETAILED EXAMINATION

Kanban systems embody the principles of pull-based production, where replenishment actions are initiated in response to actual consumption rather than forecasts. This methodology aligns inventory levels closely with real-time demand, thereby reducing waste, minimizing holding costs, and enhancing system responsiveness. To comprehensively understand the functionality and benefits of Kanban systems, it is essential to dissect the standard Kanban feedback loop step by step. Each component of the loop plays a critical role in maintaining inventory equilibrium and ensuring seamless supply chain operations.

5.1.1.1 Consumption and Inventory Depletion

The feedback loop commences with the consumption of inventory at the point of use, typically a production workstation or retail outlet. As products are consumed, the inventory level $I(t)$ at time t diminishes. This depletion is mathematically represented by the inventory balance equation:

$$\frac{dI(t)}{dt} = Q(t) - D(t),$$

where:

1. $I(t)$ is the inventory level at time t,
2. $Q(t)$ is the replenishment rate (control action) at time t,
3. $D(t)$ is the demand rate at time t.

In a discrete-time setting, this equation can be expressed as:

$$I(t+1) = I(t) + Q(t) - D(t).$$

DOI: 10.1201/9781003627906-5

This equation captures the net effect of replenishment and consumption on the inventory level, serving as the foundation for the feedback loop.

5.1.1.2 Triggering the Kanban Signal

When the inventory level $I(t)$ falls to or below a predefined threshold $I_{\text{threshold}}$, a Kanban signal $K(t)$ is activated. This signal acts as a prompt for initiating replenishment actions. The Kanban signal can be modeled as a binary function:

$$K(t) = \begin{cases} 1 & \text{if } I(t) \leq I_{\text{threshold}}, \\ 0 & \text{otherwise.} \end{cases}$$

Here, $K(t) = 1$ indicates that replenishment is required, while $K(t) = 0$ signifies that current inventory levels are sufficient. The threshold $I_{\text{threshold}}$ is strategically set based on factors such as lead time τ, demand variability, and desired service levels.

5.1.1.3 Replenishment Order Placement

Upon activation of the Kanban signal $K(t) = 1$, a replenishment order $Q(t)$ is placed. The replenishment quantity is typically standardized to a fixed amount $Q_{\max}$, representing the maximum replenishment capacity or a predetermined batch size. This can be mathematically defined as:

$$Q(t) = K(t) \cdot Q_{\max}.$$

This formulation ensures that replenishment actions are only initiated when necessary, aligning supply with actual demand and preventing overstocking.

5.1.1.4 Lead Time and Order Fulfillment

Replenishment orders are subject to lead times τ, representing the delay between placing an order and receiving the goods. During this period, the inventory continues to be consumed, and the system must account for this delay to maintain stability. The impact of lead time on the inventory dynamics is incorporated into the inventory balance equation as follows:

$$I(t+\tau) = I(t) + Q(t) - D(t).$$

Here, $I(t+\tau)$ denotes the inventory level after the lead time τ. Accurately modeling lead time is crucial for predicting inventory levels and preventing stockouts or overstock situations.

5.2 INTRODUCTION TO KANBAN SYSTEMS

Kanban, a Japanese term meaning "visual signal" or "card," is a scheduling system originally developed in the Toyota Production System (TPS) to enhance manufacturing efficiency. Kanban systems embody the principles of pull-based production, where replenishment actions are triggered by actual consumption rather than forecasts. This approach minimizes inventory levels, reduces waste, and enhances responsiveness to demand fluctuations.

In logistics, Kanban systems can be modeled as dynamic feedback mechanisms where inventory signals dictate replenishment actions. The fundamental components of a Kanban system include:

1. **Kanban Cards:** Physical or digital signals indicating the need for replenishment.
2. **Workstations:** Points in the supply chain where inventory is consumed or produced.
3. **Replenishment Processes:** Actions initiated in response to Kanban signals to restore inventory levels.

Mathematically, Kanban can be represented by a discrete-time feedback system where the inventory level at a workstation i is a function of the preceding inventory state and the triggered replenishment action:

$$I_i(t+1) = I_i(t) + Q_i(t) - D_i(t)$$

where $Q_i(t)$ is the replenishment quantity determined by the Kanban signal.

5.2.1 MODELING KANBAN AS A DYNAMIC FEEDBACK SYSTEM

Modeling Kanban systems requires capturing the interplay between inventory signals, replenishment actions, and demand patterns. A comprehensive model integrates discrete events (Kanban triggers) with continuous inventory dynamics. Consider a simplified Kanban system with the following dynamics:

$$I(t+1) = I(t) + Q(t) - D(t)$$

Here, the replenishment action $Q(t)$ is governed by the Kanban signal $K(t)$, which is a function of the inventory level:

$$K(t) = f(I(t))$$

For a binary Kanban signal:

$$K(t) = \begin{cases} 1 & \text{if } I(t) \leq I_{\text{threshold}} \\ 0 & \text{otherwise} \end{cases}$$

The replenishment quantity $Q(t)$ can be defined as:

$$Q(t) = K(t) \cdot Q_{\max}$$

where $Q_{\max}$ is the maximum allowable replenishment per Kanban cycle. Substituting into the inventory dynamics:

$$I(t+1) = I(t) + K(t) \cdot Q_{\max} - D(t)$$

This discrete-time model encapsulates the feedback nature of Kanban, where replenishment is directly tied to inventory states, fostering a responsive and efficient control system.

5.3 IMPLICATIONS FOR RELIABILITY AND SYSTEM EFFICIENCY

Incorporating feedback loops and Kanban systems into inventory control profoundly impacts system reliability and efficiency. The dynamic adjustment of replenishment actions based on real-time inventory signals enhances the system's ability to adapt to demand variability and supply chain disruptions. Key implications include:

5.3.1 ENHANCED RESPONSIVENESS

Enhanced responsiveness is a critical attribute of effective inventory control systems, particularly within the context of Kanban methodologies. Feedback mechanisms empower these systems to swiftly react to inventory deviations, thereby minimizing the risks of stockouts and excess inventory. This subsection delves into the mathematical underpinnings that facilitate enhanced responsiveness in Kanban systems, exploring how dynamic adjustments to replenishment actions align inventory levels with actual consumption patterns. Through rigorous mathematical analysis, we demonstrate how Kanban systems foster a lean and agile supply chain.

5.3.1.1 Quantifying Responsiveness

Responsiveness in inventory systems can be quantified using several performance metrics, such as **response time** and **recovery rate**. These metrics assess how quickly and effectively a system returns to equilibrium following a disturbance.

Response time T_r is defined as the time taken for the inventory level $I(t)$ to reach a certain percentage (e.g., 90%) of the target inventory level I_{target} after a deviation occurs. Mathematically, if a deviation ΔI occurs at time $t = t_0$, the response time is the smallest T_r such that:

$$I(t_0 + T_r) \geq I_{\text{target}} - \alpha \Delta I,$$

where α is a predefined tolerance level (e.g., $\alpha = 0.1$ for 90% recovery).

5.3.1.1.1 Recovery Rate

The recovery rate $R(t)$ quantifies the speed at which the inventory level approaches the target:

$$R(t) = \frac{I(t) - I_{\min}}{I_{\text{target}} - I_{\min}},$$

where $I_{\min}$ is the minimum acceptable inventory level. A higher recovery rate indicates a more responsive system.

5.3.1.2 Mathematical Modeling of Dynamic Replenishment

The core of enhanced responsiveness lies in the ability of the Kanban system to dynamically adjust replenishment actions based on real-time inventory levels. This dynamic adjustment can be modeled using difference equations in a discrete-time framework or differential equations in a continuous-time framework.

In a discrete-time Kanban system, where time advances in fixed intervals Δt (e.g., days), the inventory update equation incorporating dynamic replenishment is:

$$I(t+1) = I(t) + Q(t-\tau) - D(t),$$

where:

- $I(t)$: Inventory level at time t.
- $Q(t-\tau)$: Replenishment quantity ordered at time $t-\tau$, arriving after lead time τ.
- $D(t)$: Demand rate at time t.

The replenishment quantity $Q(t)$ is governed by the Kanban signal $K(t)$:

$$Q(t) = K(t) \cdot Q_{\max},$$

$$K(t) = \begin{cases} 1 & \text{if } I(t) \leq I_{\text{threshold}}, \\ 0 & \text{otherwise.} \end{cases}$$

This formulation ensures that replenishment actions are triggered only when inventory levels breach the threshold, allowing the system to respond promptly to consumption patterns.

Alternatively, in a continuous-time setting, the inventory dynamics are described by:

$$\frac{dI(t)}{dt} = Q(t) - D(t),$$

where $Q(t)$ is now a continuous replenishment rate. Incorporating a PID controller for $Q(t)$, the control law is:

$$Q(t) = k_p e(t) + k_i \int_0^t e(\tau)\, d\tau + k_d \frac{de(t)}{dt},$$

with $e(t) = I_{\text{target}} - I(t)$.

5.3.1.3 Transfer Function Analysis and Responsiveness

Analyzing the system's transfer function provides insights into its dynamic response characteristics, including responsiveness. The transfer function relates the Laplace transform of the output $I(s)$ to the input $D(s)$.

In the Laplace domain, the continuous-time transfer function $G_{ID}(s)$ is:

$$G_{ID}(s) = \frac{I(s)}{D(s)} = \frac{-s}{(1+k_d)s^2 + k_p s + k_i}.$$

The poles of $G_{ID}(s)$ are the roots of the characteristic equation:

$$(1+k_d)s^2+k_ps+k_i=0.$$

The location of these poles in the complex plane dictates the system's transient response. Poles with larger negative real parts result in quicker damping and faster response times, enhancing responsiveness.

5.3.1.4 Dynamic Adjustment of Replenishment Actions

Kanban systems dynamically adjust replenishment based on real-time inventory levels, which is mathematically modeled through the feedback mechanism. This dynamic adjustment enhances responsiveness by ensuring that replenishment actions are timely and proportional to consumption rates.

The replenishment rate $Q(t)$ is directly influenced by the error signal $e(t) = I_{\text{target}} - I(t)$:

$$Q(t)=k_pe(t)+k_i\int_0^t e(\tau)\,d\tau+k_d\frac{de(t)}{dt}.$$

In the context of Kanban, this translates to:

$$Q(t)=K(t)\cdot Q_{\max}.$$

Here, $K(t)$ acts as a simplified proportional controller, where k_p is implicitly set by $Q_{\max}$. The dynamic nature of $K(t)$ ensures that replenishment actions respond promptly to inventory deviations.

The responsiveness of the system is influenced by the gains k_p, k_i, and k_d. Higher proportional gains lead to quicker reactions to deviations, while integral and derivative gains fine-tune the response to prevent overshooting and oscillations. Mathematically, optimizing these gains involves balancing the trade-off between speed and stability.

5.3.1.5 Simulation of Responsiveness in Kanban Systems

Simulating Kanban systems allows for empirical analysis of responsiveness under varying conditions. Consider a discrete-time simulation with the following parameters:

1. Inventory Threshold: $I_{\text{threshold}} = 80$ units
2. Replenishment Quantity: $Q_{\max} = 100$ units
3. Lead Time: $\tau = 2$ days
4. Demand Rate: $D(t) = 20$ units/day

Simulation steps:

1. Initialize inventory: $I(0) = 100$ units.
2. For each day $t = 0, 1, 2, \ldots, T$:
   a. Consume inventory: $I(t) = I(t) - D(t)$.
   b. Check threshold: If $I(t) \leq I_{\text{threshold}}$, set $K(t) = 1$; else, $K(t) = 0$.

c. Place replenishment order: $Q(t) = K(t) \cdot Q_{\max}$.
d. Receive replenishment after lead time: $I(t+\tau) = I(t) + Q(t)$.

The simulation reveals how quickly the system reacts to inventory depletion:

1. Initial Consumption: Inventory decreases from 100 to 80 units on day 0.
2. Signal Triggering: On day 0, $I(0) = 80 \leq 80$, so $K(0) = 1$ and $Q(0) = 100$.
3. Replenishment Arrival: On day 2, $I(2) = I(2) + Q(0) = 60 + 100 = 160$.
4. Subsequent Consumption: Inventory decreases from 160 to 140 on day 2.
5. Continuous Adjustment: The system maintains inventory around the threshold, dynamically adjusting replenishment actions in response to consumption patterns.

This iterative process demonstrates the Kanban system's ability to maintain inventory levels close to the desired setpoint, highlighting its enhanced responsiveness.

5.3.1.6 Mathematical Optimization of Responsiveness

Optimizing responsiveness involves fine-tuning the Kanban system parameters to achieve desired performance metrics. This optimization can be framed as minimizing response time T_r while ensuring stability and minimizing costs.

Define an objective function that captures the trade-off between response time and cost:

$$J = \alpha T_r + \beta C,$$

where:

1. T_r: Response time.
2. C: Total inventory-related costs (holding, shortage, replenishment).
3. α, β: Weighting factors balancing the importance of responsiveness and cost.

Ensure that the system remains stable and operational:

1. $I(t) \geq I_{\min}$: Prevent stockouts.
2. $I(t) \leq I_{\max}$: Avoid excessive inventory.
3. $Q(t) \leq Q_{\max}$: Adhere to replenishment capacity.

Employ optimization algorithms such as:

1. **Linear Programming (LP):** For systems with linear dynamics and cost functions.
2. **Genetic Algorithms (GA):** For complex, non-linear optimization landscapes.
3. **Gradient Descent Methods:** For continuous optimization problems.

Minimize the objective function J subject to the constraints:

$$\min_{I(t),Q(t)} \quad \alpha T_r + \beta \left(\sum_{t=1}^{T} (hI(t) + p\max(0, I_{\text{threshold}} - I(t)) + cQ(t)) \right),$$

subject to:

$$I(t+1) = I(t) + Q(t-\tau) - D(t),$$
$$I(t) \geq I_{\min}, \quad I(t) \leq I_{\max},$$
$$Q(t) \leq Q_{\max},$$

where h, p, and c are cost coefficients for holding, shortage, and replenishment, respectively.

Solving this optimization problem yields the optimal replenishment policy that balances responsiveness with cost efficiency.

5.3.1.7 Impact of Demand Variability on Responsiveness

Demand variability significantly influences the responsiveness of Kanban systems. Higher variability necessitates more frequent replenishment actions to maintain inventory levels within desired bounds.

Assume demand $D(t)$ follows a stochastic process:

$$D(t) \sim \mathcal{N}(\mu, \sigma^2),$$

where μ is the mean demand and σ^2 is the variance.

Increased variance σ^2 leads to greater deviations from the target inventory level I_{target}, necessitating quicker and more frequent replenishment actions to restore equilibrium, thereby enhancing system responsiveness.

To accommodate demand variability, Kanban systems can incorporate safety stocks and adaptive replenishment policies:

$$I_{\text{threshold}} = I_{\text{target}} - S,$$

where S is the safety stock, determined based on demand variability:

$$S = z\sigma\sqrt{\tau},$$

with z being the z-score corresponding to the desired service level.

This ensures that the system remains responsive to demand fluctuations while maintaining inventory levels within acceptable limits.

5.3.1.8 Mathematical Insights into Responsiveness

Enhanced responsiveness in Kanban systems is achieved through the interplay of dynamic replenishment actions and real-time inventory monitoring. Mathematical analysis reveals that:

1. Feedback Gain Influence: Higher replenishment quantities $Q_{\max}$ effectively increase the feedback gain K_f, enabling quicker restoration of inventory levels post-depletion.

2. Lead Time Mitigation: Minimizing lead time τ reduces the delay between signal triggering and replenishment arrival, enhancing the system's ability to respond promptly to inventory deviations.
3. Threshold Calibration: Optimal calibration of $I_{\text{threshold}}$ balances the trade-off between responsiveness and inventory holding costs. Lower thresholds increase responsiveness but may lead to more frequent replenishments.
4. Demand Forecast Integration: Incorporating demand forecasts into the Kanban signal function can preemptively adjust replenishment actions, further enhancing system responsiveness.

5.3.1.9 Conclusion

Enhanced responsiveness in Kanban systems is fundamentally rooted in the mathematical relationship between inventory levels, replenishment actions, and demand patterns. Through dynamic feedback mechanisms, Kanban systems can swiftly adjust replenishment quantities in response to real-time inventory deviations, ensuring that inventory levels remain closely aligned with actual consumption. Mathematical modeling and analysis provide the tools necessary to optimize these feedback loops, balancing responsiveness with cost efficiency to foster a lean and agile supply chain.

By rigorously defining and quantifying responsiveness metrics, modeling dynamic replenishment actions, and employing optimization techniques, logisticians can design Kanban systems that not only respond promptly to inventory changes but also maintain stability and cost-effectiveness. This mathematical foundation is essential for developing robust inventory control strategies that meet the demands of modern, complex, and dynamic supply chains.

5.3.2 IMPROVED RELIABILITY

Reliability in inventory management pertains to the system's ability to consistently meet demand without interruptions or deviations from expected performance levels. In the context of Kanban systems, feedback loops play a pivotal role in enhancing reliability by ensuring continuous monitoring and adjustment of inventory levels. This subsection explores the mathematical frameworks and models that underpin the reliability improvements facilitated by Kanban feedback mechanisms. We examine how these models account for demand variability, supply uncertainties, and system robustness, thereby minimizing the impact of unforeseen demand spikes or supply delays.

5.3.2.1 Quantitative Metrics of Reliability

Reliability can be quantified using various metrics that assess the system's performance in maintaining service levels and operational stability. Key metrics include **Service Level**, **Fill Rate**, and **Probability of Stockouts**.

Service Level (SL) is defined as the probability of fulfilling customer demand without encountering a stockout. Mathematically, it is expressed as:

$$SL = P(D(t) \leq I(t)),$$

where $D(t)$ is the demand at time t and $I(t)$ is the inventory level at time t. A higher service level indicates greater reliability in meeting customer demand.

Fill Rate (FR) measures the proportion of customer demand that is met directly from inventory without backordering. It is calculated as:

$$FR = \frac{\sum_{t=1}^{T} \min(D(t), I(t))}{\sum_{t=1}^{T} D(t)},$$

where T is the total time period under consideration. Fill Rate provides insight into the system's effectiveness in satisfying demand.

The Probability of Stockouts (P_{stockout}) quantifies the likelihood that demand will exceed inventory levels:

$$P_{\text{stockout}} = P(D(t) > I(t)).$$

Minimizing P_{stockout} is crucial for enhancing system reliability.

2. Safety Stock Calculation

Safety stock acts as a buffer against demand variability and supply uncertainties, ensuring that inventory levels remain sufficient to meet unexpected demand surges or supply delays. The mathematical determination of safety stock is fundamental to improving reliability.

Assuming that demand $D(t)$ during lead time τ follows a normal distribution:

$$D(t) \sim \mathcal{N}(\mu_D, \sigma_D^2),$$

where μ_D is the mean demand and σ_D^2 is the variance over the lead time τ.

To achieve a desired service level SL, safety stock S can be calculated using the inverse of the standard normal distribution:

$$S = z_\alpha \sigma_D,$$

where z_α is the z-score corresponding to the desired service level $SL = 1 - \alpha$.

The inventory threshold $I_{\text{threshold}}$ is adjusted to include safety stock:

$$I_{\text{threshold}} = \mu_D + S.$$

This adjustment ensures that replenishment is triggered before inventory levels fall below the safety stock, thereby reducing the probability of stockouts.

5.3.2.2 Demand and Supply Variability Modeling

Reliability improvements hinge on accurately modeling the variability in both demand and supply. Stochastic processes and probability distributions are employed to capture these uncertainties.

Demand variability is characterized by its mean μ_D and standard deviation σ_D. Common models include:

1. **Normal Distribution:** Suitable for continuous demand variables with symmetric variability.
$$D(t) \sim \mathcal{N}(\mu_D, \sigma_D^2).$$
2. **Poisson Distribution:** Appropriate for discrete demand events occurring independently over time.
$$D(t) \sim \text{Poisson}(\lambda_D),$$
where λ_D is the average demand rate.

Supply variability encompasses delays and disruptions in replenishment. It can be modeled using delay distributions or failure probabilities:

1. **Deterministic Lead Time:** A fixed lead time τ.
2. **Stochastic Lead Time:** Lead time τ follows a probability distribution, such as exponential or normal.
$$\tau \sim \mathcal{N}(\mu_\tau, \sigma_\tau^2).$$

5.3.2.3 Reliability Optimization through Feedback Control

Optimizing reliability involves adjusting system parameters to minimize the Probability of Stockouts while maintaining cost efficiency. Feedback control in Kanban systems facilitates this optimization by dynamically responding to inventory deviations.

Formulate the reliability optimization as:

$$\min_{I_{\text{threshold}}, Q_{\max}} \quad P_{\text{stockout}} = P(D(t) > I(t)),$$

$$\text{subject to:} \quad \text{Cost Constraints},$$

where cost constraints include holding costs, shortage costs, and replenishment costs.

To incorporate cost constraints, use the Lagrangian multiplier method:

$$\mathcal{L}(I_{\text{threshold}}, Q_{\max}, \lambda) = P_{\text{stockout}} + \lambda (C_{\text{holding}} + C_{\text{shortage}} + C_{\text{replenishment}} - C_{\text{budget}}).$$

Solve the optimization problem by finding the stationary points of $\mathcal{L}$.

Employ numerical optimization methods such as:

1. **Gradient Descent:** Iteratively update parameters in the direction of the steepest descent.

2. **Genetic Algorithms:** Utilize evolutionary strategies to explore the parameter space.
3. **Simulated Annealing:** Search for the global minimum by probabilistically accepting worse solutions early on.

5.3.2.4 Reliability under Demand and Supply Shocks

Real-world systems are subject to sudden changes in demand or supply conditions, known as shocks. Modeling and mitigating these shocks is essential for maintaining reliability.

Demand shocks are abrupt increases or decreases in demand. Mathematically, model a demand shock at time $t = t_s$ as:

$$D(t) = \begin{cases} D_0 & \text{if } t \neq t_s, \\ D_0 + \Delta D & \text{if } t = t_s, \end{cases}$$

where D_0 is the baseline demand and ΔD is the demand surge.

Supply shocks involve unexpected delays or failures in replenishment. Model a supply shock by introducing an additional delay $\Delta\tau$ at time $t = t_s$:

$$\tau(t) = \begin{cases} \tau_0 & \text{if } t \neq t_s, \\ \tau_0 + \Delta\tau & \text{if } t = t_s, \end{cases}$$

where τ_0 is the standard lead time.

Evaluate how shocks affect reliability metrics:

$$SL(t_s) = P(D(t_s) \leq I(t_s)),$$

$$FR(t_s) = \frac{\min(D(t_s), I(t_s))}{D(t_s)}.$$

Analyze the degradation in service levels and fill rates due to shocks, and adjust safety stock and replenishment policies accordingly to restore reliability.

5.3.2.5 Robustness of Kanban Systems

Robustness refers to the system's ability to maintain reliability despite uncertainties and disturbances. Mathematical models assess robustness by evaluating performance under varying demand and supply conditions.

Apply robust control theory to design Kanban systems that remain reliable under parameter variations:

$$\min_{I_{\text{threshold}}, Q_{\max}} \max_{\delta D, \delta\tau} P_{\text{stockout}}(I(t), D(t) + \delta D, \tau + \delta\tau),$$

$$\text{subject to: Cost Constraints},$$

where δD and $\delta\tau$ represent perturbations in demand and lead time, respectively.

Utilize H_∞ control methods to ensure that the system's worst-case performance is optimized:

$$\min_{I_{\text{threshold}}, Q_{\max}} \quad \|G_{ID}(s)\|_\infty,$$

where $G_{ID}(s)$ is the transfer function from demand to inventory. Minimizing the H_∞ norm ensures bounded responses to all possible disturbances.

5.3.2.6 Mathematical Insights into Reliability Improvement

Enhanced reliability through feedback loops in Kanban systems is underpinned by several mathematical principles and models:

1. Markov Chains. Model the inventory states as a Markov chain, where each state represents a specific inventory level. Transition probabilities between states are determined by demand and replenishment actions. Analyzing the steady-state probabilities provides insights into the long-term reliability of the system.

$$P(I(t+1) = j | I(t) = i) = P(D(t) = i + Q(t) - j).$$

2. Queueing Theory. Apply queueing theory to model the inventory system as a service system, where demand arrivals are analogous to customer arrivals and replenishments act as service completions. The reliability can be assessed by evaluating metrics such as queue length (inventory level) and waiting time (replenishment lead time).
3. Stochastic differential equations. Use stochastic differential equations (SDEs) to model continuous-time inventory dynamics under uncertainty:

$$dI(t) = (Q(t) - \mu_D)dt + \sigma_D dW(t),$$

where $W(t)$ is a Wiener process representing demand fluctuations.
4. Robustness metrics. Employ robustness metrics from control theory to evaluate the system's resilience to parameter variations and external disturbances. Metrics such as gain and phase margins quantify the system's ability to maintain reliability under changing conditions.

$$\text{Gain Margin (GM)} = \frac{1}{|G(j\omega_c)|},$$
$$\text{Phase Margin (PM)} = 180^\circ + \angle G(j\omega_c),$$

where ω_c is the critical frequency at which the phase of $G(j\omega_c)$ is -180°.
5. Optimization of Safety Stock. Optimize safety stock S by minimizing the total cost function:

$$C_{\text{total}} = C_{\text{holding}} + C_{\text{shortage}},$$
$$C_{\text{holding}} = hS,$$
$$C_{\text{shortage}} = p(1 - SL).$$

Solve for S that minimizes C_{total} while achieving the desired service level SL.

$$S^* = \frac{p}{h} z_\alpha \sigma_D \sqrt{\tau}.$$

5.3.2.7 Practical Implementation Considerations

Implementing mathematically optimized reliability enhancements in Kanban systems requires careful consideration of practical factors:

1. Accurate demand forecasting and reliable measurement of lead times are essential for precise safety stock calculations and system reliability.
2. As supply chains grow in complexity, the mathematical models must scale accordingly, accommodating multiple products, varied demand patterns, and interconnected supply nodes.
3. Integrate real-time data analytics and monitoring tools to continuously assess inventory levels and demand patterns, enabling timely adjustments to replenishment actions.
4. Conduct thorough cost-benefit analyses to balance the investments in safety stock and system robustness against the savings from reduced stockouts and enhanced service levels.

5.3.2.8 Conclusion

Mathematical modeling and analysis are indispensable for enhancing the reliability of Kanban systems through feedback loops. By quantifying reliability metrics, optimizing safety stock, and employing robust control strategies, Kanban systems can effectively mitigate the impacts of demand variability and supply uncertainties. The integration of advanced mathematical techniques such as Markov chains, queueing theory, and stochastic processes provides a comprehensive framework for designing reliable and resilient inventory control systems. As supply chains continue to evolve, the mathematical rigor underpinning Kanban feedback mechanisms ensures that these systems remain robust, efficient, and capable of sustaining high service levels in dynamic and uncertain environments.

Through a detailed mathematical exploration, this subsection has elucidated the mechanisms by which feedback loops enhance the reliability of Kanban systems. These insights empower logisticians and engineers to design and optimize inventory control strategies that uphold operational stability and consistently meet customer demand, thereby reinforcing the foundational role of Kanban in modern, lean supply chain management.

5.3.3 PULL-BASED VS. PUSH-BASED SYSTEMS

Inventory management systems can broadly be categorized into push-based and pull-based approaches. Understanding the mathematical distinctions between these systems is crucial for designing inventory control mechanisms that align with organizational objectives and operational dynamics. This subsection provides a rigorous mathematical comparison of push-based and pull-based (Kanban) systems, highlighting their respective control strategies, response to demand variability, and implications for supply chain efficiency.

5.3.3.1 Control Strategies

Push-based and pull-based systems employ fundamentally different control strategies for inventory replenishment, which can be formalized through their respective replenishment functions.

In push-based systems, replenishment actions are dictated by forecasts and production schedules. The replenishment quantity $Q(t)$ at time t is determined by a function of the forecasted demand $D(t)$:

$$Q(t) = f_{\text{forecast}}(D(t)),$$

where f_{forecast} is typically a linear or non-linear function that aggregates forecasted demand over a planning horizon.

Conversely, pull-based systems like Kanban initiate replenishment based on actual consumption signals. The replenishment quantity $Q(t)$ is a function of the current inventory level $I(t)$:

$$Q(t) = f_{\text{signal}}(I(t)),$$

where f_{signal} is often a threshold-based function that triggers replenishment when inventory falls below a specified level.

5.3.3.2 Mathematical Modeling of Replenishment Functions

To elucidate the operational differences, we model both replenishment strategies using discrete-time systems.

A common push-based replenishment strategy is the fixed-order quantity (FOQ) method, where replenishment is based on forecasted demand over the lead time τ. Mathematically, this can be expressed as:

$$Q(t) = k \cdot \hat{D}(t),$$

where:

1. k is a scaling factor reflecting order multiples or safety stock considerations,
2. $\hat{D}(t)$ is the forecasted demand over the lead time period.

Alternatively, push-based systems may employ more sophisticated forecasting models, such as exponential smoothing or ARIMA (AutoRegressive Integrated Moving Average), to predict future demand:

$$\hat{D}(t) = \alpha D(t-1) + \beta \hat{D}(t-1) + \gamma \varepsilon(t-1),$$

where α, β, and γ are smoothing parameters, and $\varepsilon(t-1)$ represents forecast errors.

In pull-based systems, replenishment is directly tied to actual inventory levels. A standard Kanban replenishment function employs a binary trigger:

$$f_{\text{signal}}(I(t)) = \begin{cases} Q_{\max} & \text{if } I(t) \leq I_{\text{threshold}}, \\ 0 & \text{otherwise}, \end{cases}$$

where:

1. $Q_{\max}$ is the fixed replenishment quantity,
2. $I_{\text{threshold}}$ is the inventory level at which replenishment is triggered.

This approach ensures that replenishment actions are reactive to real-time inventory states, thereby reducing reliance on demand forecasts.

5.3.3.3 System Dynamics and Stability Analysis

The dynamic behavior and stability of push-based and pull-based systems can be analyzed using difference equations and control theory principles.

For a push-based system with fixed-order quantity replenishment, the inventory update equation is:

$$I(t+1) = I(t) + Q(t) - D(t),$$
$$Q(t) = k \cdot \hat{D}(t).$$

Assuming a steady-state forecast where $\hat{D}(t) = D(t) = D$, the system simplifies to:

$$I(t+1) = I(t) + kD - D = I(t) + (k-1)D.$$

The system reaches equilibrium when $k = 1$, leading to:

$$I^* = I^* + 0 \implies I^* \text{ is constant.}$$

However, deviations from equilibrium ($k \neq 1$) can lead to unbounded inventory levels or perpetual shortages, indicating potential instability:

$$I(t+1) - I(t) = (k-1)D.$$

To maintain stability, k must be carefully calibrated, typically requiring $k = 1$, which may not be feasible in dynamic demand environments.

In contrast, the pull-based Kanban system employs a feedback loop where replenishment is contingent upon inventory levels:

$$I(t+1) = I(t) + Q(t-\tau) - D(t),$$

$$Q(t) = \begin{cases} Q_{\max} & \text{if } I(t) \leq I_{\text{threshold}}, \\ 0 & \text{otherwise.} \end{cases}$$

Assuming deterministic demand $D(t) = D$ and lead time τ, the system's stability can be analyzed by examining the inventory dynamics over time. The introduction of feedback through the Kanban signal inherently adjusts replenishment actions based on real-time inventory states, promoting oscillatory behavior around the threshold $I_{\text{threshold}}$.

For the pull-based system to remain stable, the replenishment quantity must satisfy:

$$Q_{\max} \geq D \cdot \tau,$$

ensuring that replenishment actions can compensate for demand over the lead time period. Mathematically, this condition prevents the system from depleting inventory levels to zero, thereby avoiding stockouts.

The transient response of push-based systems is heavily dependent on the accuracy of demand forecasts. Inaccurate forecasts can lead to significant inventory oscillations or divergences. Conversely, pull-based systems inherently adapt to actual demand, resulting in a more controlled and predictable transient response. The feedback mechanism in Kanban systems dampens the impact of demand variability, enhancing system robustness.

5.3.3.4 Response to Demand Variability

The ability to respond to demand variability is a key differentiator between push-based and pull-based systems.

Push-based systems are susceptible to forecast errors. Mathematically, if the actual demand $D(t)$ deviates from the forecast $\hat{D}(t)$, the inventory level evolves as:

$$I(t+1) = I(t) + k\hat{D}(t) - D(t).$$

Forecast errors $\varepsilon(t) = \hat{D}(t) - D(t)$ propagate through the system, potentially causing inventory levels to either surge or deplete:

$$I(t+1) = I(t) + kD(t) + k\varepsilon(t) - D(t) = I(t) + (k-1)D(t) + k\varepsilon(t).$$

Persistent forecast inaccuracies can lead to unbounded inventory growth ($k > 1$) or persistent shortages ($k < 1$), undermining system reliability.

Pull-based systems inherently mitigate the impact of demand variability by basing replenishment on actual inventory states:

$$I(t+1) = I(t) + Q(t-\tau) - D(t).$$

Since replenishment is triggered by real-time consumption, the system dynamically adjusts to fluctuations in demand without relying on forecasts. Mathematically, this can be seen as:

$$Q(t) = \begin{cases} Q_{\max} & \text{if } I(t) \leq I_{\text{threshold}}, \\ 0 & \text{otherwise.} \end{cases}$$

This feedback-driven approach ensures that replenishment actions are responsive to actual demand patterns, enhancing system adaptability and reducing the risk of stockouts or overstocking.

5.3.3.5 Cost Implications

The choice between push-based and pull-based systems has significant cost implications, which can be analyzed mathematically through cost functions.

Push-based systems incur costs related to:

1. **Holding Costs:** Associated with maintaining inventory levels $I(t)$.
2. **Shortage Costs:** Arising from unmet demand when $I(t) < D(t)$.
3. **Replenishment Costs:** Linked to order placement and transportation.

The total cost C_{push} can be modeled as:

$$C_{\text{push}} = \sum_{t=1}^{T} \left(hI(t) + p\max(0, D(t) - I(t)) + cQ(t)\right),$$

where h, p, and c are the holding, shortage, and replenishment cost coefficients, respectively.

Pull-based systems also incur holding, shortage, and replenishment costs, but the dynamic adjustment of replenishment actions typically results in lower holding and shortage costs:

$$C_{\text{pull}} = \sum_{t=1}^{T} \left(hI(t) + p\max(0, D(t) - I(t)) + cQ(t)\right).$$

Mathematically, since $Q(t)$ in pull-based systems is contingent on actual inventory states, the system tends to maintain inventory levels closer to the threshold, reducing excess holding and minimizing shortage occurrences.

By formulating and comparing the cost functions C_{push} and C_{pull}, we can evaluate the economic efficiency of each system under varying demand conditions. Push-based systems may exhibit higher total costs due to forecast inaccuracies, while pull-based systems benefit from cost savings through adaptive replenishment.

5.3.3.6 Mathematical Optimization of Replenishment Strategies

Optimizing replenishment strategies involves determining the optimal parameters that minimize total costs while maintaining desired service levels.

Define the objective function to minimize total costs:

$$\min_{Q_{\max}, I_{\text{threshold}}} \quad C_{\text{total}} = \sum_{t=1}^{T} \left(hI(t) + p\max(0, D(t) - I(t)) + cQ(t)\right),$$

subject to:

$$Q(t) = \begin{cases} f_{\text{forecast}}(D(t)) & \text{push-based,} \\ f_{\text{signal}}(I(t)) & \text{pull-based,} \end{cases}$$

$$I(t+1) = I(t) + Q(t - \tau) - D(t).$$

Incorporate constraints to ensure system stability and operational feasibility:

1. $Q(t) \geq 0$: Non-negativity of replenishment quantities.
2. $I(t) \geq 0$: Prevent negative inventory levels.
3. $I(t) \leq I_{\max}$: Avoid excessive inventory accumulation.

5.3.3.7 Robustness Against Forecast Errors

Push-based systems are inherently vulnerable to forecast inaccuracies, which can be modeled as stochastic perturbations in the replenishment function:

$$Q(t) = k \cdot \hat{D}(t) + \varepsilon(t),$$

where $\varepsilon(t)$ represents forecast errors. These errors propagate through the inventory dynamics, potentially destabilizing the system:

$$I(t+1) = I(t) + k \cdot \hat{D}(t) + \varepsilon(t) - D(t).$$

Pull-based systems mitigate this vulnerability by basing replenishment on actual inventory levels, effectively reducing the impact of forecast errors:

$$I(t+1) = I(t) + K(t-\tau) \cdot Q_{\max} - D(t).$$

Thus, mathematically, pull-based systems offer enhanced robustness against uncertainties in demand forecasting.

5.3.3.8 Conclusion

The mathematical comparison of push-based and pull-based (Kanban) systems underscores the inherent advantages of feedback-driven replenishment strategies in modern inventory management. Push-based systems, reliant on demand forecasts, are susceptible to inefficiencies and instabilities arising from forecast inaccuracies. In contrast, pull-based systems leverage real-time consumption signals to dynamically adjust replenishment actions, fostering greater responsiveness, reliability, and cost efficiency.

Through rigorous mathematical modeling and analysis, it becomes evident that pull-based systems offer a more robust and adaptable framework for inventory control, particularly in environments characterized by demand variability and supply uncertainties. The feedback mechanisms inherent in Kanban systems enable seamless alignment between supply and actual demand, thereby enhancing overall supply chain performance and operational sustainability.

As supply chains continue to evolve in complexity and globalization, the mathematical foundations of pull-based inventory control provide essential insights for designing resilient and efficient logistics systems. This comparative analysis equips logisticians and engineers with the knowledge to implement optimal replenishment strategies that balance cost, service levels, and system stability, ensuring sustained competitive advantage in dynamic market landscapes.

5.3.4 MODELING KANBAN AS A FEEDBACK CONTROLLER

To model Kanban as a feedback controller, consider the following components:

1. **Sensor:** Monitors the inventory level $I(t)$.
2. **Controller:** Processes the inventory signal $K(t)$ based on $I(t)$.
3. **Actuator:** Initiates replenishment actions $Q(t)$ in response to $K(t)$.

The feedback loop can be mathematically modeled as:

$$K(t) = f(I(t))$$

$$Q(t) = g(K(t))$$

$$I(t+1) = I(t) + Q(t) - D(t)$$

For a binary Kanban system:

$$K(t) = \begin{cases} 1 & \text{if } I(t) \leq I_{\text{threshold}} \\ 0 & \text{otherwise} \end{cases}$$

$$Q(t) = K(t) \cdot Q_{\text{max}}$$

Thus, the inventory dynamics under Kanban control are:

$$I(t+1) = I(t) + K(t) \cdot Q_{\text{max}} - D(t)$$

5.3.5 STABILITY AND TRANSIENT RESPONSE IN KANBAN SYSTEMS

The stability of Kanban-controlled systems hinges on the interplay between replenishment rates, demand variability, and inventory thresholds. Analyzing the system's transient response involves examining how quickly and smoothly the inventory level converges to the desired setpoint without excessive oscillations or overshooting.

At equilibrium, $I(t+1) = I(t) = I^*$, and assuming constant demand $D(t) = D$, the equilibrium inventory level I^* satisfies:

$$I^* = I^* + K(t) \cdot Q_{\text{max}} - D$$

Simplifying:

$$K(t) \cdot Q_{\text{max}} = D$$

Given $K(t) = 1$ when $I(t) \leq I_{\text{threshold}}$, the equilibrium condition ensures that the replenishment rate matches the demand rate when inventory reaches the threshold, maintaining a balance between supply and consumption.

The transient behavior can be characterized by the system's response to perturbations or changes in demand. For instance, consider a step increase in demand $D(t)$, prompting the feedback controller to adjust replenishment actions accordingly. The resulting inventory level trajectory depends on the system's parameters, such as $Q_{\max}$ and $I_{\text{threshold}}$.

$$I(t+1) = I(t) + K(t) \cdot Q_{\max} - D(t)$$

The dynamics can be further analyzed by iterating the difference equation, examining the conditions under which the system converges to the new equilibrium without oscillations or instability.

5.3.6 FEEDBACK LOOPS IN KANBAN SYSTEMS: MATHEMATICAL INSIGHTS

To gain a deeper understanding of feedback loops within Kanban systems, it is essential to explore their mathematical underpinnings. This involves analyzing the system's stability, sensitivity to parameter variations, and robustness against demand fluctuations.

5.3.7 STABILITY ANALYSIS

Stability in feedback-controlled systems is determined by the system's ability to return to equilibrium after a disturbance. For Kanban systems, this involves ensuring that inventory levels stabilize around the setpoint without persistent oscillations or divergence.

Consider the discrete-time difference equation:

$$I(t+1) = I(t) + K(t) \cdot Q_{\max} - D(t)$$

Assuming constant demand $D(t) = D$ and a binary Kanban signal:

$$K(t) = \begin{cases} 1 & \text{if } I(t) \leq I_{\text{threshold}} \\ 0 & \text{otherwise} \end{cases}$$

The system's behavior can be categorized based on the relationship between $Q_{\max}$ and D:

1. **Under-Replenishment:** If $Q_{\max} < D$, the system cannot sustain the desired inventory level, leading to perpetual stockouts.
2. **Balanced Replenishment:** If $Q_{\max} = D$, the system maintains inventory levels at the threshold, providing just enough replenishment to meet demand.
3. **Over-Replenishment:** If $Q_{\max} > D$, the system can overshoot the inventory threshold, potentially triggering replenishment cycles and oscillatory behavior.

For stability, the system must satisfy:

$$Q_{\max} \geq D$$

This ensures that the replenishment rate can at least match the demand rate, preventing inventory depletion over time.

5.3.8 SENSITIVITY TO PARAMETER VARIATIONS

The responsiveness and stability of Kanban systems are highly sensitive to parameter settings, particularly the inventory threshold $I_{\text{threshold}}$ and the replenishment quantity Q_{max}.

The threshold determines the point at which replenishment is triggered. A lower $I_{\text{threshold}}$ reduces the inventory holding costs but increases the frequency of replenishment actions. Conversely, a higher threshold enhances buffer stock levels, improving service reliability but elevating holding costs.

Mathematically, the choice of $I_{\text{threshold}}$ affects the system's dynamic response. Optimal threshold selection involves balancing the trade-off between holding costs and service levels.

The replenishment quantity directly influences the system's ability to absorb demand fluctuations. Larger Q_{max} values enhance the system's capacity to handle demand spikes but can induce oscillatory behavior if not carefully managed.

5.4 DESIGNING RELIABLE AND EFFICIENT FEEDBACK-CONTROLLED KANBAN SYSTEMS

The design of reliable and efficient feedback-controlled Kanban systems necessitates a holistic approach that integrates mathematical modeling, controller design, and performance optimization. Key design principles include:

5.4.1 ACCURATE SYSTEM MODELING

Developing precise mathematical models of inventory dynamics, demand patterns, and replenishment actions is foundational for effective controller design. Accurate models enable the prediction of system behavior and facilitate the tuning of feedback parameters to achieve desired performance.

5.4.2 CONTROLLER TUNING AND OPTIMIZATION

Optimizing controller parameters (k_p, k_i, k_d) is essential for balancing responsiveness and stability. Techniques such as optimization-based tuning, genetic algorithms, and machine learning can be employed to identify optimal parameter settings that minimize cost functions while satisfying operational constraints.

5.4.3 INCORPORATING SAFETY STOCKS AND BUFFER ZONES

Incorporating safety stocks and buffer zones within the Kanban framework enhances system reliability by providing additional inventory buffers against demand variability and supply chain disruptions. Mathematically, safety stocks can be integrated into the inventory threshold:

$$I_{\text{threshold}} = I_{\text{desired}} - S$$

where S is the safety stock level.

5.4.4 CONTINUOUS MONITORING AND FEEDBACK

Implementing continuous monitoring systems that track inventory levels, demand fluctuations, and replenishment actions in real-time ensures that feedback loops operate effectively. Integrating advanced data analytics and real-time tracking technologies enhances the system's ability to respond dynamically to changing conditions.

5.5 FUTURE DIRECTIONS, EMERGING TRENDS AND CONCLUSION

Designing Kanban systems with scalability and flexibility in mind allows them to adapt to varying supply chain complexities and operational scales. Modular design principles and distributed control architectures facilitate the expansion and modification of Kanban systems to accommodate growth and evolving logistical requirements. As logistics systems continue to evolve, feedback-controlled Kanban systems must adapt to incorporate emerging technologies and methodologies. Future directions include:

1. Integration with Industry 4.0 Technologies. The advent of Industry 4.0 brings forth technologies such as the Internet of Things (IoT), artificial intelligence (AI), and big data analytics. Integrating these technologies with Kanban systems enhances their intelligence and autonomy, enabling predictive replenishment and advanced decision-making capabilities.
2. IoT-Enabled Kanban Systems. IoT devices can provide real-time data on inventory levels, environmental conditions, and supply chain statuses, enabling more precise and timely feedback control actions within Kanban systems.
3. Machine Learning and Predictive Analytics. Machine learning algorithms can analyze historical inventory and demand data to predict future trends, enabling Kanban systems to anticipate demand fluctuations and adjust replenishment actions proactively.

By leveraging predictive analytics, Kanban systems can transition from purely reactive to semi-predictive control strategies, enhancing their ability to maintain optimal inventory levels under uncertain conditions.

Feedback loops and Kanban systems represent a synergistic integration of control theory and inventory management principles, fostering responsive, reliable, and efficient logistics operations. By modeling inventory signals and logistical concepts as mathematical functions, logisticians can design and optimize Kanban systems that dynamically adjust to demand fluctuations and supply chain disruptions. The mathematical rigor underpinning feedback-controlled Kanban systems facilitates their analysis, stability assessment, and performance optimization, ensuring that supply chains remain agile and cost-effective in an increasingly complex and dynamic global marketplace.

6 Network Science in Logistics

6.1 INTRODUCTION

In the contemporary landscape of global commerce, the intricacies of logistics and supply chain management have grown exponentially in complexity. As organizations strive to achieve operational excellence, the integration of advanced analytical frameworks becomes indispensable. Among these, network science emerges as a pivotal discipline, offering profound insights into the structural and dynamic aspects of logistics networks. This chapter, titled *Network Science in Logistics*, serves as an exposition for bachelor students, elucidating the application of various network topologies and network analysis techniques to evaluate connectivity, robustness, and sustainability within supply chains.

6.1.1 THE INTERSECTION OF NETWORK SCIENCE AND LOGISTICS

Network science, at its core, examines the interplay between nodes (entities) and edges (connections) within a system. In the context of logistics, nodes represent critical components such as suppliers, manufacturers, distribution centers, retailers, and customers, while edges symbolize the flow of goods, information, and financial transactions between these entities. By modeling supply chains as networks, logisticians and engineers can leverage mathematical tools to analyze and optimize the flow of resources, mitigate risks, and enhance overall system resilience.

6.1.2 MATHEMATICAL FOUNDATIONS OF LOGISTICS NETWORKS

To effectively apply network science to logistics, it is imperative to establish a robust mathematical framework. A logistics network can be formally represented as a graph $G = (V, E)$, where V denotes the set of nodes and E the set of edges. Each node $v_i \in V$ embodies a logistical entity, while each edge $e_{ij} \in E$ captures the relationship or interaction between nodes v_i and v_j.

One of the fundamental representations of a graph is its adjacency matrix A, defined as:

$$A = [a_{ij}]_{n \times n},$$

where $a_{ij} = 1$ if there is an edge from node v_i to node v_j, and $a_{ij} = 0$ otherwise. For weighted networks, a_{ij} can represent the capacity, cost, or other relevant metrics associated with the connection.

The degree of a node, k_i, signifies the number of edges incident to node v_i. The degree distribution $P(k)$ provides the probability that a randomly selected node has

DOI: 10.1201/9781003627906-6

degree k. Understanding the degree distribution is crucial for assessing network properties such as connectivity and vulnerability.

6.1.3 NETWORK TOPOLOGIES IN LOGISTICS

Different network topologies impart distinct characteristics to logistics networks, influencing their efficiency, robustness, and scalability. This subsection explores several prominent network configurations and their applicability to logistics.

6.1.3.1 Random Networks

Random networks, exemplified by the Erdoes-Rényi model $G(n, p)$, where n is the number of nodes and p the probability of edge formation between any pair of nodes, provide a baseline for understanding network behavior. In logistics, random networks can model scenarios where connections between entities are formed without specific constraints or strategic planning. While simple to analyze, random networks often lack the structural properties necessary for efficient supply chain operations, such as redundancy and targeted connectivity.

6.1.3.2 Small-World Networks

Small-world networks, characterized by high clustering coefficients and short average path lengths, are modeled by the Watts-Strogatz framework. In logistics, small-world topologies can facilitate rapid information dissemination and efficient resource allocation across the supply chain. The presence of shortcuts in the network reduces the number of steps required for goods and information to traverse from suppliers to consumers, thereby enhancing operational speed and flexibility.

6.1.3.3 Scale-Free Networks

Scale-free networks follow a power-law degree distribution, $P(k) \sim k^{-\gamma}$, indicating the presence of hubs—nodes with exceptionally high degrees. The Barabási-Albert model captures the growth dynamics of such networks through preferential attachment. In logistics, scale-free topologies can represent centralized distribution centers or key suppliers that serve as critical nodes within the supply chain. While the presence of hubs can enhance efficiency by consolidating resources, it also introduces vulnerability, as the failure of a hub can have cascading effects on the entire network.

6.1.3.4 Hierarchical Networks

Hierarchical networks exhibit a multi-tiered structure, reflecting the hierarchical nature of many supply chains. These networks consist of layers, such as raw material suppliers, manufacturers, distributors, and retailers, with interconnections predominantly occurring within and between adjacent layers. Hierarchical topologies facilitate organized resource allocation and streamline management processes, although

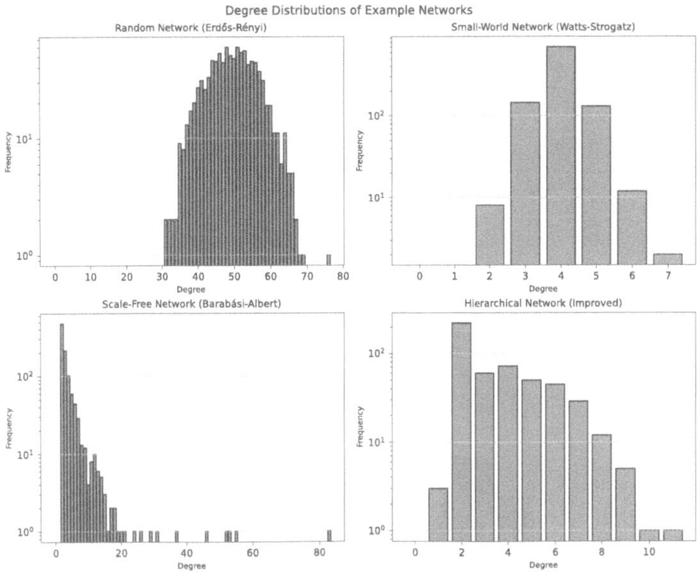

Figure 6.1 Degre distribution of exemplary networks.

they may be less resilient to disruptions compared to more decentralized network structures.

As shown in Figure 6.1, we depict the degree distributions of exemplary networks.

6.1.4 SPECTRAL PROPERTIES OF LOGISTICS NETWORKS

The spectral properties of a network, derived from the eigenvalues and eigenvectors of matrices associated with the graph (such as the adjacency matrix or Laplacian matrix), offer deep insights into the network's structural and dynamical characteristics. In logistics, spectral analysis can be employed to evaluate connectivity, detect bottlenecks, and assess robustness.

6.1.4.1 Laplacian Matrix and Eigenvalues

The Laplacian matrix L of a graph, exemplarily shown in Figure 6.2, is defined as:

$$L = D - A,$$

where D is the degree matrix (a diagonal matrix where $D_{ii} = k_i$) and A is the adjacency matrix. The eigenvalues of the Laplacian matrix, particularly the second smallest eigenvalue known as the algebraic connectivity or Fiedler value λ_2, are indicative of the network's connectivity. A larger λ_2 implies a more robust and well-connected network, whereas a smaller λ_2 suggests vulnerability to disconnection.

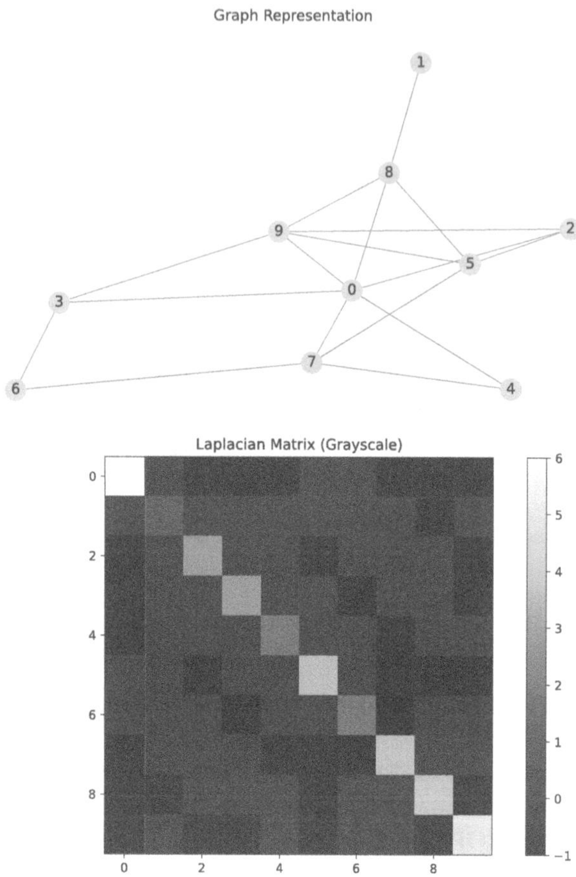

Figure 6.2 Graph and Laplacian matrix of exemplary network.

The corresponding eigenvector to λ_2, known as the Fiedler vector, provides information about the network's community structure and can be utilized for partitioning the network into clusters. In logistics, identifying such clusters can aid in optimizing regional distribution strategies and enhancing localized supply chain resilience.

6.1.4.2 Spectral Graph Theory in Logistics

Spectral graph theory bridges the gap between graph structures and their spectral properties, enabling the application of linear algebra techniques to network analysis. In logistics, spectral graph theory can be leveraged to:

1. **Optimize Routing:** By analyzing the spectral gap (the difference between the first and second smallest eigenvalues), logisticians can assess the efficiency of routing protocols and identify potential areas for improvement.

2. **Enhance Robustness:** Networks with favorable spectral properties are less susceptible to random failures and targeted attacks. Spectral resilience metrics can guide the design of robust supply chains capable of withstanding disruptions.
3. **Facilitate Network Synchronization:** Ensuring synchronized operations across the supply chain can be mathematically modeled and optimized using spectral properties, promoting seamless coordination between different logistical entities.

6.1.5 CONNECTIVITY AND ROBUSTNESS IN SUPPLY CHAINS

Connectivity and robustness are paramount for the resilience and efficiency of supply chains. Network science provides the tools to quantify and enhance these attributes through rigorous mathematical analysis.

6.1.5.1 Connectivity Metrics

Connectivity metrics measure the extent to which nodes in a network are interconnected. Key metrics include:

1. **Average Path Length:** The average number of steps along the shortest paths for all possible pairs of network nodes. It is mathematically defined as:

$$L = \frac{1}{n(n-1)} \sum_{i \neq j} d(v_i, v_j)$$

   where:

   a. n is the total number of nodes in the network ($|V|$),
   b. $d(v_i, v_j)$ is the shortest path distance between node v_i and node v_j.
   Shorter path lengths facilitate faster communication and resource distribution across the network.
2. **Clustering Coefficient:** Measures the degree to which nodes in a network tend to form clusters. The local clustering coefficient for a node v_i is given by:

$$C(v_i) = \frac{2e_i}{k_i(k_i - 1)}$$

   where:

   a. e_i is the number of edges between the neighbors of v_i,
   b. k_i is the degree of node v_i (the number of neighbors of v_i).
   The overall clustering coefficient of the network is the average clustering coefficient across all nodes:

$$C = \frac{1}{n} \sum_{i=1}^{n} C(v_i)$$

   High clustering can indicate localized supply chain efficiencies but may also suggest redundancy.

3. **Degree Distribution:** Represents the probability distribution of node degrees across the network. The degree k_i of a node v_i is the number of edges connected to it. The degree distribution $P(k)$ is defined as:

$$P(k) = \frac{\text{Number of nodes with degree } k}{\text{Total number of nodes } n}$$

Degree distribution influences the network's vulnerability and efficiency. For example:

a. In random networks, $P(k)$ often follows a Poisson distribution.
b. In scale-free networks, $P(k) \sim k^{-\gamma}$, where γ is a constant.

6.1.5.2 Robustness Analysis

Robustness refers to the network's ability to maintain functionality in the face of node or edge failures. Mathematical models assess robustness by simulating various failure scenarios and analyzing the resulting impact on network connectivity and performance.

Percolation theory studies the movement and filtering of fluids through porous materials but has been adapted to analyze the robustness of networks. In logistics, percolation models can simulate the impact of random failures (random node removals) or targeted attacks (removal of high-degree nodes) on the supply chain's connectivity.

$$P_{\text{percolation}} = f(p),$$

where p is the probability of node removal. The percolation threshold p_c marks the critical point beyond which the network disintegrates into disconnected clusters.

Mathematically, introducing redundancy by adding alternative pathways or backup suppliers can be modeled as increasing the network's edge density or enhancing node connectivity. This approach reduces the percolation threshold, thereby increasing the network's robustness.

6.1.5.3 Resilience Optimization

Resilience optimization involves designing supply chains that can swiftly recover from disruptions. Mathematical optimization techniques can be employed to determine optimal network configurations that maximize resilience while minimizing costs.

Formulate the resilience optimization as:

$$\max_{G} \quad R(G),$$

$$\text{subject to:} \quad C(G) \leq C_{\max},$$

where $R(G)$ is a resilience metric (e.g., algebraic connectivity), $C(G)$ represents the total cost associated with network configuration G, and $C_{\max}$ is the budget constraint.

Employ combinatorial optimization techniques such as integer programming, genetic algorithms, and simulated annealing to explore the vast space of possible network configurations and identify those that offer optimal resilience.

6.1.6 SUSTAINABLE MANUFACTURING AND NETWORK DESIGN

Sustainable manufacturing integrates environmental considerations into the production process, aiming to minimize ecological footprints while maintaining economic viability. Network science contributes to sustainable manufacturing by optimizing the supply chain network to reduce resource consumption, emissions, and waste.

6.1.6.1 Green Network Topologies

Designing green network topologies involves creating supply chain configurations that minimize environmental impact. Mathematical models can optimize the placement of manufacturing facilities, distribution centers, and transportation routes to achieve sustainability goals.

Carbon Footprint Minimization:

Formulate the optimization problem as:

$$\min_{G} \sum_{e \in E} c_e x_e,$$

where c_e represents the carbon emissions associated with edge e, and x_e is a binary variable indicating the presence or absence of edge e. Subject to connectivity and operational constraints, this formulation seeks to minimize the total carbon footprint of the supply chain network.

Employ graph-theoretic algorithms to identify energy-efficient routing paths that reduce transportation-related energy consumption. Metrics such as edge betweenness centrality can highlight critical pathways where energy savings can be maximized through optimization.

6.1.6.2 Circular Supply Chains

Circular supply chains aim to create closed-loop systems where products and materials are reused, refurbished, or recycled, thereby reducing waste and conserving resources. Mathematical models facilitate the design of such systems by optimizing the flow of materials and minimizing environmental impact.

Model the circular supply chain as a directed graph with additional edges representing reverse flows for recycling and remanufacturing. Optimize the flow variables to maximize material reuse while minimizing transportation and processing costs:

$$\max \sum_{e \in E_{\text{reverse}}} f_e,$$

$$\text{subject to:} \quad \text{Flow Conservation Constraints},$$

where E_{reverse} denotes the set of reverse edges facilitating circular flows.

Integrate life cycle assessment (LCA) metrics into network optimization models to evaluate the environmental impact of different network configurations and identify sustainable practices.

6.1.7 INVENTORY SIGNALS AS MATHEMATICAL FUNCTIONS IN NETWORKED LOGISTICS

Inventory signals, pivotal in inventory control mechanisms such as Kanban systems, can be mathematically modeled as functions that interact with the network's structural properties. Treating inventory signals as mathematical functions allows for the integration of inventory dynamics with network analysis, enabling a comprehensive evaluation of supply chain performance.

6.1.7.1 Functional Representation of Inventory Signals

Define inventory signals $S(v_i,t)$ as functions dependent on node-specific inventory levels $I(v_i,t)$ and possibly other factors such as demand rates $D(v_i,t)$ and lead times $\tau(v_i,t)$:

$$S(v_i,t) = f(I(v_i,t), D(v_i,t), \tau(v_i,t)),$$

where f encapsulates the decision rules for triggering replenishment actions based on real-time inventory states.

6.1.7.2 Integration with Network Dynamics

In a networked logistics system, the inventory signals at individual nodes influence and are influenced by the broader network structure. Mathematically, this interaction can be modeled through coupled difference or differential equations that account for the flow of goods and information across the network:

$$I(v_i,t+1) = I(v_i,t) + \sum_{v_j \in N(v_i)} Q(v_j \to v_i,t) - D(v_i,t),$$

$$Q(v_j \to v_i,t) = g(S(v_j,t), A(v_j,v_i)),$$

where $N(v_i)$ denotes the neighbors of node v_i, $Q(v_j \to v_i,t)$ represents the replenishment flow from node v_j to node v_i, g is a function determining the replenishment quantity based on the inventory signal $S(v_j,t)$ and the adjacency $A(v_j,v_i)$.

6.1.7.3 Dynamic Network Optimization

Leverage dynamic optimization techniques to continuously adjust network parameters in response to inventory signals, enhancing the supply chain's adaptability and efficiency. This involves solving time-dependent optimization problems that account for evolving inventory states and network configurations:

$$\min_{Q(t)} \sum_{t=1}^{T} \sum_{e \in E} C_e Q_e(t),$$

$$\text{subject to:} \quad I(v_i, t+1) = I(v_i, t) + \sum_{v_j \in N(v_i)} Q(v_j \to v_i, t) - D(v_i, t),$$

$$S(v_i, t) = f(I(v_i, t), D(v_i, t), \tau(v_i, t)),$$

$$Q(v_j \to v_i, t) \leq Q_{\max}(e),$$

where C_e denotes the cost associated with edge e, and $Q_{\max}(e)$ is the maximum replenishment capacity for edge e.

6.1.8 SUSTAINABILITY CONSIDERATIONS IN NETWORK DESIGN

Incorporating sustainability into network design requires a multi-faceted approach that balances economic efficiency with environmental stewardship. Mathematical models play a crucial role in quantifying and optimizing sustainability metrics within logistics networks.

6.1.8.1 Multi-Objective Optimization

Formulate the network design problem as a multi-objective optimization problem, simultaneously minimizing costs and environmental impacts:

$$\min_{G} \left(\sum_{e \in E} C_e Q_e, \sum_{e \in E} E_e Q_e \right),$$

$$\text{subject to:} \quad \text{Connectivity and Operational Constraints},$$

where E_e represents the environmental cost (e.g., carbon emissions) associated with edge e.

Identify Pareto-efficient solutions where no objective can be improved without degrading another. This allows logisticians to select network configurations that offer the best trade-offs between cost and sustainability.

6.1.8.2 Green Routing Algorithms

Develop and implement routing algorithms that prioritize environmentally friendly pathways. Mathematically, this involves defining cost functions that incorporate both economic and environmental factors:

$$\min_{R} \sum_{e \in R} (C_e + \alpha E_e),$$

where α is a weighting factor that balances economic and environmental costs.

Compare traditional shortest path algorithms with green path algorithms to evaluate the trade-offs between operational efficiency and sustainability.

6.1.8.3 Life Cycle Assessment (LCA) Integration

Integrate Life Cycle Assessment (LCA) methodologies into network analysis to evaluate the environmental impact of logistics operations from cradle to grave. Mathematically, model the cumulative environmental impact $LCA(G)$ as:

$$LCA(G) = \sum_{e \in E} \sum_{t=1}^{T} \phi_e Q_e(t),$$

where ϕ_e represents the environmental impact factor for edge e.

6.1.9 CASE STUDIES AND PRACTICAL APPLICATIONS

To bridge theoretical concepts with real-world applications, this chapter incorporates case studies that demonstrate the application of network science principles to enhance logistics operations. These case studies illustrate how mathematical models and network analysis techniques can be employed to solve complex logistical challenges, optimize supply chain performance, and promote sustainability.

6.1.9.1 Optimizing Distribution Networks

Analyze a case study involving the optimization of a distribution network for a multinational corporation. Utilize spectral graph theory to identify critical nodes and edges, and apply network robustness metrics to ensure supply chain resilience against disruptions.

Mathematical approach:

1. Construct the distribution network as a graph $G = (V, E)$.
2. Calculate the Laplacian matrix L and determine the algebraic connectivity λ_2.
3. Identify vulnerable nodes using the Fiedler vector.
4. Optimize the network by reinforcing connections or adding redundant paths to enhance λ_2.

6.1.9.2 Sustainable Supply Chain Design

Examine a case study focused on designing a sustainable supply chain for an electronics manufacturer. Employ multi-objective optimization to balance cost efficiency with carbon footprint reduction.

Mathematical Approach:

1. Define objective functions for cost C and environmental impact E.
2. Utilize linear programming to minimize C and E subject to supply chain constraints.
3. Analyze Pareto frontiers to identify optimal trade-offs.
4. Implement green routing algorithms to reduce transportation emissions.

6.1.9.3 Resilient Network Design Under Uncertainty

Investigate a case study on enhancing the resilience of a pharmaceutical supply chain amidst global disruptions. Apply percolation theory and robustness metrics to evaluate network stability under various failure scenarios.

Mathematical Approach:

1. Model the pharmaceutical supply chain as a graph $G = (V, E)$.
2. Simulate random and targeted node removals to assess network robustness.
3. Calculate the percolation threshold p_c and evaluate the network's resilience.
4. Optimize network topology to increase p_c, thereby enhancing resilience.

6.1.10 FUTURE DIRECTIONS AND EMERGING TRENDS

As logistics systems continue to evolve, the integration of network science with emerging technologies presents new opportunities for optimization and innovation. Future research directions include:

1. **Dynamic Network Modeling:** Developing time-dependent models that capture the temporal evolution of supply chains.
2. **Integration with Internet of Things (IoT):** Leveraging real-time data from IoT devices to enhance network monitoring and control.
3. **Machine Learning and Predictive Analytics:** Applying machine learning algorithms to predict network failures and optimize supply chain performance.
4. **Blockchain for Supply Chain Transparency:** Utilizing blockchain technology to enhance traceability and trust within logistics networks.

6.1.11 CONCLUSION

The application of network science to logistics offers a mathematically rigorous framework for analyzing and optimizing supply chain networks. By modeling logistics as networks, leveraging spectral graph theory, and employing robust optimization techniques, logisticians can enhance connectivity, robustness, and sustainability. This chapter has outlined the foundational concepts and mathematical tools essential for understanding and applying network science to logistics, setting the stage for more detailed explorations of specific network topologies and their implications for supply chain performance.

As the global economy becomes increasingly interconnected and supply chains more intricate, the insights derived from network science will be instrumental in designing resilient, efficient, and sustainable logistics systems. Bachelor students equipped with this knowledge will be well-prepared to tackle the challenges of modern logistics and contribute to the advancement of supply chain management through the application of sophisticated mathematical models and network analysis techniques.

6.2 OUTLINE OF THE CHAPTER

Following this introduction, the chapter proceeds with an in-depth exploration of various network topologies and their mathematical properties as applied to logistics. Each section delves into specific aspects of network science, providing detailed mathematical formulations and analyses relevant to supply chain connectivity and robustness. The subsequent sections include:

1. **Spectral Properties of Logistics Networks:** Analyzing eigenvalues and eigenvectors of the Laplacian matrix to assess connectivity and identify critical nodes using the Fiedler vector.
2. **Graph Configurations and Degree Distributions:** Examining scale-free, small-world, random, and hierarchical network models to understand their impact on supply chain efficiency and vulnerability.
3. **Connectivity and Robustness Metrics:** Defining and calculating metrics such as algebraic connectivity, clustering coefficients, and percolation thresholds to evaluate network resilience.
4. **Optimization Techniques for Network Design:** Applying linear programming, dynamic programming, and heuristic algorithms to design optimal logistics networks that balance cost, efficiency, and sustainability.
5. **Case Studies and Applications:** Presenting real-world scenarios where network science principles have been successfully implemented to enhance logistics operations.
6. **Emerging Trends and Future Research Directions:** Exploring the integration of network science with advanced technologies such as IoT, machine learning, and blockchain to drive innovation in supply chain management.

Each section is enriched with mathematical derivations, illustrative examples, and practical applications, ensuring a comprehensive understanding of how network science can be harnessed to optimize logistics and supply chain operations. By the end of the chapter, students will have acquired the mathematical expertise and analytical skills necessary to apply network science principles to real-world logistical challenges, fostering the development of efficient, resilient, and sustainable supply chains.

6.3 MATHEMATICAL PRELIMINARIES

Before delving into the application of network science to logistics, it is essential to establish the mathematical foundations that underpin network analysis. This section introduces key concepts and mathematical tools, including graph theory fundamentals, spectral graph theory, and optimization techniques, which will be utilized throughout the chapter.

6.3.1 GRAPH THEORY FUNDAMENTALS

Graph theory provides the structural framework for modeling logistics networks. Key definitions include:

1. **Graph:** A graph $G = (V, E)$ consists of a set of vertices V and a set of edges E.
2. **Directed vs. Undirected Graphs:** Directed graphs have edges with a specific direction $(v_i \to v_j)$, while undirected graphs have bidirectional edges $(v_i - v_j)$.
3. **Weighted Graphs:** Graphs where edges have associated weights w_{ij}, representing metrics such as distance, cost, or capacity.
4. **Subgraphs and Clusters:** Subgraphs $G' = (V', E')$ represent subsets of the main graph, useful for identifying clusters or modules within the supply chain.

6.3.2 SPECTRAL GRAPH THEORY

Spectral graph theory explores the properties of graphs through the eigenvalues and eigenvectors of matrices associated with the graph. The Laplacian matrix L and adjacency matrix A are central to this analysis.

The Laplacian matrix L is defined as:

$$L = D - A,$$

where D is the degree matrix and A the adjacency matrix. The eigenvalues of L provide insights into the graph's connectivity and robustness.

The eigenvalues λ of L satisfy:

$$L\mathbf{x} = \lambda \mathbf{x},$$

where $\mathbf{x}$ is an eigenvector. The spectrum of L, particularly the second smallest eigenvalue λ_2 (algebraic connectivity), is indicative of the graph's resilience to node or edge removals.

The eigenvector corresponding to λ_2, known as the Fiedler vector, is instrumental in graph partitioning and community detection, aiding in the identification of critical nodes within the supply chain network.

6.3.3 OPTIMIZATION TECHNIQUES

Optimization is integral to designing efficient logistics networks. This subsection outlines fundamental optimization methods relevant to network design and analysis.

Linear Programming (LP) involves optimizing a linear objective function subject to linear equality and inequality constraints. In logistics, LP can be applied to minimize transportation costs, balance supply and demand, and optimize resource allocation.

$$\min_{\mathbf{x}} \quad \mathbf{c}^\top \mathbf{x},$$

$$\text{subject to:} \quad \mathbf{A}\mathbf{x} \leq \mathbf{b}, \quad \mathbf{x} \geq 0.$$

Integer Programming (IP) extends LP by requiring some or all decision variables to take integer values, suitable for discrete decisions such as facility location and network design.

Dynamic Programming (DP) solves complex problems by breaking them down into simpler subproblems, applicable to multi-period inventory optimization and network flow problems.

For large-scale and non-linear optimization problems, heuristic algorithms (e.g., greedy algorithms) and metaheuristic algorithms (e.g., genetic algorithms, simulated annealing) provide approximate solutions when exact methods are computationally infeasible.

6.3.4 NETWORK ROBUSTNESS METRICS

Quantifying network robustness involves assessing the network's ability to maintain functionality under various stress conditions. Key metrics include:

1. **Algebraic Connectivity (λ_2):** Measures the graph's connectivity, with higher values indicating greater robustness.
2. **Network Resilience Index:** A composite metric that evaluates the network's ability to recover from disruptions.
3. **Redundancy Index:** Quantifies the extent of redundant pathways within the network.

6.3.5 INTEGRATING INVENTORY SIGNALS INTO NETWORK MODELS

Inventory signals, essential for real-time inventory control, can be seamlessly integrated into network models by treating them as mathematical functions that influence and are influenced by the network's structural properties.

Define inventory signals $S(v_i,t)$ at node v_i and time t as:

$$S(v_i,t) = f(I(v_i,t), D(v_i,t), \tau(v_i,t)),$$

where f encapsulates the logic for triggering replenishment based on current inventory levels $I(v_i,t)$, demand rates $D(v_i,t)$, and lead times $\tau(v_i,t)$.

Model the interaction between network topology and inventory dynamics through coupled equations:

$$I(v_i,t+1) = I(v_i,t) + \sum_{v_j \in N(v_i)} Q(v_j \to v_i,t) - D(v_i,t),$$

$$Q(v_j \to v_i,t) = g(S(v_j,t), A(v_j,v_i)),$$

where $Q(v_j \to v_i, t)$ represents the replenishment flow from node v_j to node v_i, and g is a function determining the replenishment quantity based on the inventory signal and adjacency matrix $A(v_j, v_i)$.

6.3.6 CONCLUSION

This introductory section has established the foundational concepts and mathematical frameworks essential for understanding the application of network science to logistics. By modeling supply chains as complex networks and employing mathematical tools such as spectral graph theory and optimization techniques, logisticians and engineers can analyze and enhance the connectivity, robustness, and sustainability of supply chains. The subsequent sections will delve deeper into specific network topologies, spectral properties, and optimization strategies, providing a comprehensive exploration of how network science can be harnessed to optimize logistics and supply chain operations in an increasingly complex and dynamic global environment.

6.4 GRAPH SPECTRAL PROPERTIES AND SUPPLY CHAIN CONNECTIVITY

Building upon the mathematical preliminaries outlined in the introduction, this section delves into the spectral properties of graphs and their implications for supply chain connectivity and robustness. By analyzing the eigenvalues and eigenvectors of graph-associated matrices, we can derive critical insights into the structural integrity and operational efficiency of logistics networks.

6.4.1 LAPLACIAN MATRIX EIGENVALUES AND CONNECTIVITY

The Laplacian matrix L plays a central role in spectral graph theory, providing a bridge between graph structure and its dynamic properties.

6.4.1.1 Algebraic Connectivity (λ_2)

The second smallest eigenvalue of the Laplacian matrix, known as the algebraic connectivity λ_2, is a pivotal metric in assessing network robustness. Formally, λ_2 is defined as:

$$\lambda_2 = \min_{\mathbf{x} \perp \mathbf{1}} \frac{\mathbf{x}^\top L \mathbf{x}}{\mathbf{x}^\top \mathbf{x}},$$

where $\mathbf{1}$ is the all-ones vector, and $\mathbf{x}$ is orthogonal to $\mathbf{1}$.

A higher λ_2 signifies a more tightly connected network, less susceptible to disconnections upon node or edge removals. Conversely, a lower λ_2 indicates potential vulnerabilities and the presence of bottlenecks within the supply chain.

6.4.1.2 Fiedler Vector and Network Partitioning

The eigenvector corresponding to λ_2, known as the Fiedler vector $\mathbf{v}_2$, provides valuable information about the network's community structure and potential points of failure.

Utilizing the Fiedler vector, one can partition the graph into communities or clusters by identifying nodes with similar vector components. Mathematically, nodes can be divided based on the sign or magnitude of their corresponding entries in $\mathbf{v}_2$:

$$\text{Cluster 1: } v_i \text{ if } (\mathbf{v}_2)_i \geq 0,$$

$$\text{Cluster 2: } v_i \text{ if } (\mathbf{v}_2)_i < 0.$$

This partitioning aids in identifying critical segments of the supply chain that require redundancy or enhanced connectivity to prevent disruptions.

6.4.2 SPECTRAL GAP AND NETWORK ROBUSTNESS

The spectral gap, defined as the difference between λ_2 and the smallest non-zero eigenvalue λ_1 (which is zero for connected graphs), provides a measure of network robustness.

$$\text{Spectral Gap} = \lambda_2 - \lambda_1.$$

A larger spectral gap indicates a more robust network with enhanced resilience against random failures and targeted attacks. In logistics, maximizing the spectral gap can be achieved by increasing the connectivity of critical nodes and ensuring the absence of weakly connected subgraphs.

6.4.3 GRAPH CONFIGURATIONS AND THEIR SPECTRAL PROPERTIES

Different network topologies exhibit distinct spectral characteristics, influencing their suitability for various logistical applications.

6.4.3.1 Scale-Free Networks

Scale-free networks, characterized by a power-law degree distribution, exhibit a few highly connected hubs and numerous nodes with low degrees. The presence of hubs significantly affects the spectral properties of the network:

1. **High λ_2:** Despite having hubs, scale-free networks often maintain relatively high algebraic connectivity due to the centrality of hubs.
2. **Vulnerability of Hubs:** The removal of hubs can dramatically decrease λ_2, highlighting the importance of protecting key nodes to preserve network integrity.

6.4.3.2 Small-World Networks

Small-world networks combine high clustering with short average path lengths, providing efficient communication and resource distribution:

1. **Moderate λ_2:** Small-world networks typically exhibit moderate algebraic connectivity, balancing local clustering with global connectivity.
2. **Enhanced Path Redundancy:** The presence of shortcuts reduces the dependence on specific nodes, thereby enhancing network robustness.

6.4.3.3 Random Networks

Random networks, with their uniform edge distribution, present different spectral characteristics:

1. **Low λ_2:** Random networks often have lower algebraic connectivity compared to scale-free or small-world networks, making them less robust against random failures.
2. **Uniform Vulnerability:** The lack of highly connected hubs results in uniform vulnerability across the network, simplifying resilience strategies but potentially limiting overall robustness.

6.4.4 DEGREE DISTRIBUTION AND SPECTRAL PROPERTIES

The degree distribution of a network profoundly influences its spectral properties and, consequently, its logistical performance.

6.4.4.1 Power-Law Distribution

Scale-free networks adhere to a power-law degree distribution:

$$P(k) \sim k^{-\gamma},$$

where γ typically ranges between 2 and 3. This distribution results in the emergence of hubs, which play a critical role in enhancing λ_2 and overall network connectivity.

6.4.4.2 Poisson Distribution

Random networks follow a Poisson degree distribution:

$$P(k) = \frac{\lambda^k e^{-\lambda}}{k!},$$

where λ is the average degree. This distribution lacks hubs, leading to lower algebraic connectivity and increased susceptibility to fragmentation upon node removals.

6.4.4.3 Exponential Distribution

Small-world networks may exhibit an exponential degree distribution in certain contexts, balancing between the uniformity of random networks and the heterogeneity of scale-free networks.

Networks with heterogeneous degree distributions (e.g., scale-free) tend to have higher λ_2 compared to homogeneous networks (e.g., random), enhancing their robustness and efficiency in logistics applications.

6.4.5 OPTIMIZATION OF NETWORK TOPOLOGIES FOR LOGISTICS

Mathematically optimizing network topologies involves adjusting the graph's structure to achieve desired performance metrics, such as enhanced connectivity, reduced path lengths, and improved robustness.

6.4.5.1 Graph Augmentation

Graph augmentation entails adding edges to the network to increase connectivity and reduce average path lengths:

$$\lambda_2(G') \geq \lambda_2(G),$$

where G' is the augmented graph. This process can be formulated as an optimization problem:

$$\max_{E'} \quad \lambda_2(G \cup E'),$$

$$\text{subject to:} \quad |E'| \leq k,$$

where E' is the set of edges to be added and k is the maximum number of additional edges.

Incorporate cost constraints to ensure that network enhancements are economically feasible:

$$\sum_{e \in E'} C_e \leq C_{\text{budget}},$$

where C_e represents the cost of adding edge e.

6.4.5.2 Edge Rewiring

Edge rewiring involves modifying existing connections to optimize network properties without increasing the total number of edges. Mathematically, this can be expressed as:

$$\min_{R} \quad \sum_{e \in R} \left| C_e^{\text{new}} - C_e^{\text{old}} \right|,$$

where R is the set of edges to be rewired, and C_e^{new}, C_e^{old} are the costs associated with the new and old edges, respectively.

Rewiring strategies aim to maximize the spectral gap λ_2 to enhance network robustness:

$$\lambda_2(G') - \lambda_2(G) \geq \delta,$$

where δ is a predefined improvement threshold.

6.4.5.3 Node Placement and Facility Location

Optimizing node placement involves determining the optimal locations for key logistical facilities (e.g., warehouses, distribution centers) to enhance network connectivity and reduce operational costs.

Define a binary decision variable x_i for each potential facility location $v_i \in V$:

$$x_i = \begin{cases} 1 & \text{if a facility is placed at } v_i, \\ 0 & \text{otherwise.} \end{cases}$$

Formulate the optimization problem to minimize total transportation costs while ensuring network connectivity:

$$\min_{\mathbf{x}} \quad \sum_{(v_i, v_j) \in E} c_{ij} x_i x_j,$$

$$\text{subject to:} \quad \text{Connectivity Constraints}, \quad \sum_{i \in V} x_i \leq K,$$

where c_{ij} represents the cost of transportation between nodes v_i and v_j, and K is the maximum number of facilities allowed.

Incorporate spectral metrics to ensure that the placement of facilities contributes to a higher algebraic connectivity λ_2:

$$\max_{\mathbf{x}} \quad \lambda_2(G(\mathbf{x})),$$

$$\text{subject to:} \quad \sum_{i \in V} x_i \leq K.$$

6.4.6 CONCLUSION

The spectral analysis of logistics networks offers a mathematically rigorous approach to evaluating and enhancing supply chain connectivity and robustness. By leveraging the eigenvalues and eigenvectors of graph-associated matrices, logisticians can gain deep insights into network vulnerabilities, optimize network topologies, and design resilient supply chains capable of withstanding disruptions. The subsequent sections will further explore these concepts, delving into specific network configurations, optimization strategies, and real-world applications that demonstrate the transformative potential of network science in logistics.

6.5 GRAPH CONFIGURATIONS AND THEIR IMPACT ON SUPPLY CHAIN EFFICIENCY

Building on the foundational understanding of spectral properties and connectivity, this section examines various graph configurations and their implications for supply chain efficiency and robustness. Each network topology—random, small-world, scale-free, and hierarchical—presents unique structural attributes that influence logistical performance. By mathematically characterizing these topologies, we can discern their suitability for different logistical contexts and identify strategies for optimizing supply chain networks.

6.5.1 RANDOM NETWORKS IN LOGISTICS

Random networks, often modeled using the Erdős-Rényi $G(n, p)$ model, serve as a baseline for understanding supply chain dynamics in the absence of structured connectivity.

6.5.1.1 Structural Properties

In a random network $G(n, p)$, each possible edge between n nodes exists with independent probability p. Key characteristics include:

- **Homogeneous Degree Distribution:** Most nodes have degrees close to the average $\lambda = p(n-1)$.
- **Low Clustering Coefficient:** The probability that two neighbors of a node are connected is low, resulting in fewer tightly-knit clusters.
- **High Average Path Length:** The typical number of steps required to traverse between any two nodes is relatively large.

6.5.1.2 Mathematical Implications for Logistics

1. **Efficiency of Resource Distribution:** High average path lengths can lead to increased transportation costs and longer lead times.
2. **Vulnerability to Failures:** The lack of redundancy and hub nodes makes random networks susceptible to fragmentation upon random node or edge failures.
3. **Optimization Challenges:** Enhancing connectivity in random networks often requires significant infrastructural investments due to the absence of naturally efficient pathways.

6.5.2 SMALL-WORLD NETWORKS AND SUPPLY CHAIN AGILITY

Small-world networks, characterized by high clustering coefficients and short average path lengths, strike a balance between regular lattices and random networks, making them highly relevant for logistics applications.

6.5.2.1 Structural Properties

Modeled by the Watts-Strogatz $G(n, k, p)$ model, small-world networks possess:

1. **High Clustering Coefficient:** Nodes tend to form tightly-knit clusters, promoting localized efficiency.

2. **Short Average Path Length:** The existence of shortcuts significantly reduces the number of steps required for traversal.
3. **Moderate Degree Heterogeneity:** Degree distribution exhibits some variability but lacks the extreme heterogeneity of scale-free networks.

6.5.2.2 Mathematical Implications for Logistics

1. **Enhanced Communication Efficiency:** Short average path lengths facilitate rapid information dissemination across the supply chain, enabling timely decision-making.
2. **Localized Resilience:** High clustering fosters localized redundancy, allowing parts of the supply chain to operate independently in the event of disruptions.
3. **Balanced Robustness:** While not as robust as scale-free networks against targeted attacks, small-world networks offer improved resilience compared to random networks.

6.5.3 SCALE-FREE NETWORKS AND CENTRALIZED SUPPLY CHAINS

Scale-free networks, distinguished by their power-law degree distributions, are emblematic of many real-world supply chains where certain nodes serve as critical hubs.

6.5.3.1 Structural Properties

The Barabási-Albert $G(n,m)$ model generates scale-free networks with:

1. **Power-Law Degree Distribution:** $P(k) \sim k^{-\gamma}$, with $\gamma \approx 3$, leading to the presence of highly connected hubs.
2. **Low Average Path Length:** Similar to small-world networks, scale-free networks exhibit short path lengths due to hub connectivity.
3. **High Clustering Coefficient:** Although not as high as small-world networks, scale-free networks maintain significant clustering.

6.5.3.2 Mathematical Implications for Logistics

1. **Efficiency of Bulk Operations:** Hubs enable the consolidation of resources, reducing overall transportation and operational costs.
2. **Centralized Control:** The presence of hubs allows for centralized decision-making, enhancing coordination and control within the supply chain.
3. **Vulnerability of Hubs:** The failure of hub nodes can lead to disproportionate disruptions, necessitating strategies for hub protection and redundancy.

6.5.4 HIERARCHICAL NETWORKS AND MULTI-TIER SUPPLY CHAINS

Hierarchical networks reflect the multi-tiered structure of many supply chains, encompassing various levels from raw material suppliers to end consumers.

6.5.4.1 Structural Properties

Hierarchical networks are characterized by:

1. **Multi-Tiered Structure:** Nodes are organized into distinct layers, each representing different functional levels within the supply chain.
2. **High Modularity:** The network comprises well-defined modules or communities, each corresponding to a specific tier or function.
3. **Controlled Connectivity:** Interconnections primarily occur between adjacent tiers, maintaining a clear hierarchical flow of resources and information.

6.5.4.2 Mathematical Implications for Logistics

1. **Structured Resource Allocation:** Hierarchical networks facilitate organized allocation of resources, ensuring efficient production and distribution processes.
2. **Scalability:** The modular nature of hierarchical networks allows for scalability, accommodating growth and expansion of the supply chain without compromising structural integrity.
3. **Potential Bottlenecks:** Centralized tiers within hierarchical networks can become bottlenecks, necessitating careful management and optimization to prevent delays and inefficiencies.

6.5.5 COMPARATIVE ANALYSIS OF NETWORK TOPOLOGIES

A comparative mathematical analysis of different network topologies reveals their respective strengths and weaknesses in logistical applications.

6.5.5.1 Connectivity and Robustness

1. **Random Networks:** Offer low robustness and efficiency, with high susceptibility to fragmentation.
2. **Small-World Networks:** Provide a balance of efficiency and moderate robustness, suitable for dynamic and flexible supply chains.
3. **Scale-Free Networks:** Excel in efficiency and centralized control but face significant vulnerabilities due to hub dependency.
4. **Hierarchical Networks:** Ensure structured resource distribution and scalability but may encounter bottlenecks and centralized points of failure.

6.5.5.2 Spectral Characteristics

Analyze how each topology influences spectral properties such as algebraic connectivity λ_2 and the spectral gap:

1. **Random Networks:** Typically have lower λ_2, indicating weaker connectivity.

2. **Small-World Networks:** Exhibit higher λ_2 due to the presence of shortcuts, enhancing overall connectivity.
3. **Scale-Free Networks:** Maintain high λ_2 through hub connectivity but are vulnerable to targeted attacks on hubs.
4. **Hierarchical Networks:** Feature varying λ_2 depending on inter-tier connectivity and redundancy.

6.5.5.3 Cost and Efficiency

Mathematically evaluate the cost-efficiency trade-offs inherent in each network topology:

1. **Random Networks:** Often incur higher costs due to inefficient routing and lack of strategic resource placement.
2. **Small-World Networks:** Achieve cost savings through reduced path lengths and enhanced communication efficiency.
3. **Scale-Free Networks:** Optimize costs via centralized hubs but may require significant investment in hub protection and redundancy.
4. **Hierarchical Networks:** Balance costs through structured resource allocation but may face inefficiencies due to potential bottlenecks.

6.5.6 MATHEMATICAL MODELING OF SUPPLY CHAIN NETWORKS

To comprehensively analyze and compare different network topologies, mathematical models must accurately represent the supply chain dynamics and interactions between nodes.

6.5.6.1 Flow Networks

Model the supply chain as a flow network $G = (V, E, c, d)$, where:

1. V: Set of nodes (suppliers, manufacturers, distributors, retailers).
2. E: Set of directed edges representing the flow of goods and information.
3. $c : E \to \mathbb{R}^+$: Capacity function denoting the maximum flow capacity of each edge.
4. $d : V \to \mathbb{R}$: Demand function specifying the required flow at each node.

The maximum flow f^* from a source node s to a sink node t in a flow network is equal to the capacity of the minimum cut $C_{\min}$ that separates s and t:

$$f^* = C_{\min}.$$

In logistics, this theorem can be applied to determine the optimal transportation routes that maximize flow efficiency and minimize bottlenecks.

6.5.6.2 Graph Laplacian and Flow Dynamics

Incorporate flow dynamics into the Laplacian framework to model the distribution and redistribution of resources within the network.

At each node v_i, the net flow must satisfy:

$$\sum_{v_j \in N(v_i)} f_{ji} - \sum_{v_j \in N(v_i)} f_{ij} = d(v_i),$$

where f_{ij} represents the flow from node v_i to node v_j.

Express the flow dynamics using the Laplacian matrix L:

$$L\mathbf{f} = \mathbf{d},$$

where **f** is the flow vector and **d** the demand vector. Solving these equations provides the optimal flow distribution that meets demand while adhering to capacity constraints.

6.5.6.3 Spectral Optimization

Leverage spectral properties to optimize flow distribution and network robustness:

1. **Enhancing λ_2:** Increase algebraic connectivity to improve flow distribution efficiency and reduce vulnerability.
2. **Minimizing Eigenvalue Spread:** Reduce the spread between eigenvalues to achieve a more uniform flow distribution and prevent congestion.

6.5.7 ADVANCED NETWORK METRICS FOR LOGISTICS

Beyond basic connectivity and spectral properties, advanced network metrics provide deeper insights into the performance and efficiency of supply chain networks.

6.5.7.1 Betweenness Centrality

Betweenness centrality $C_B(v_i)$ measures the extent to which a node v_i lies on shortest paths between other nodes:

$$C_B(v_i) = \sum_{s \neq v_i \neq t} \frac{\sigma_{st}(v_i)}{\sigma_{st}},$$

where σ_{st} is the total number of shortest paths from node s to node t, and $\sigma_{st}(v_i)$ is the number of those paths passing through v_i.

Nodes with high betweenness centrality are critical for maintaining efficient flow within the supply chain. Identifying and reinforcing such nodes can prevent bottlenecks and enhance overall network performance.

6.5.7.2 Closeness Centrality

Closeness centrality $C_C(v_i)$ quantifies how close a node v_i is to all other nodes in the network:

$$C_C(v_i) = \frac{1}{\sum_{v_j \in V} d(v_i, v_j)},$$

where $d(v_i, v_j)$ is the shortest path distance between nodes v_i and v_j.

Nodes with high closeness centrality can rapidly disseminate information and resources, enhancing the supply chain's responsiveness and agility.

6.5.7.3 Eigenvector Centrality

Eigenvector centrality $C_E(v_i)$ assesses a node's influence based on the connectivity of its neighbors:

$$C_E(v_i) = \frac{1}{\lambda} \sum_{v_j \in N(v_i)} C_E(v_j),$$

where λ is a normalization constant.

Nodes with high eigenvector centrality are not only well-connected but also connected to other influential nodes, playing a pivotal role in the network's overall efficiency and resilience.

6.5.8 APPLICATION OF NETWORK METRICS TO SUPPLY CHAIN OPTIMIZATION

Mathematically integrating network metrics into supply chain optimization frameworks enables targeted enhancements that improve connectivity, reduce vulnerabilities, and optimize resource allocation.

6.5.8.1 Flow Optimization Using Centrality Metrics

Incorporate centrality metrics into flow optimization models to prioritize resource allocation:

$$\max_{\mathbf{f}} \sum_{v_i \in V} C_E(v_i) f_i,$$

$$\text{subject to:} \quad \mathbf{Af} \leq \mathbf{c}, \quad \mathbf{f} \geq 0,$$

where f_i represents the flow through node v_i, $C_E(v_i)$ the eigenvector centrality, and $\mathbf{c}$ the capacity constraints.

This approach ensures that resources are allocated to nodes that maximize the network's overall influence and efficiency, enhancing the supply chain's performance.

6.5.8.2 Robustness Enhancement through Spectral Gap Maximization

Maximize the spectral gap to enhance network robustness:

$$\max_{G} \quad \lambda_2(G),$$

$$\text{subject to:} \quad |E| \leq E_{\max},$$

where $|E|$ is the number of edges and $E_{\max}$ is the maximum allowable edges. This optimization ensures that the network remains connected and resilient against disruptions while adhering to cost constraints.

Employ gradient-based optimization techniques or heuristic algorithms to iteratively adjust the network topology, increasing λ_2 while maintaining budgetary and operational constraints.

6.5.9 INTEGRATING INVENTORY SIGNALS WITH NETWORK CONNECTIVITY

The integration of inventory signals with network connectivity metrics facilitates a holistic approach to supply chain management, where inventory control and network design are mutually reinforcing.

6.5.9.1 Dynamic Inventory Signal Functions

Define dynamic inventory signal functions $S(v_i,t)$ that adapt based on network connectivity and operational conditions:

$$S(v_i,t) = f\left(I(v_i,t), C_B(v_i), C_C(v_i), \lambda_2(G)\right),$$

where $C_B(v_i)$ and $C_C(v_i)$ are the betweenness and closeness centralities of node v_i, respectively, and $\lambda_2(G)$ is the algebraic connectivity of the network.

Utilize $S(v_i,t)$ to implement adaptive replenishment strategies that prioritize nodes based on their centrality and the network's overall connectivity:

$$Q(v_i,t) = g\left(S(v_i,t), A(v_i)\right),$$

where $A(v_i)$ represents the adjacency matrix entries for node v_i.

This integration ensures that replenishment actions are strategically aligned with the network's structural properties, enhancing efficiency, reducing costs, and mitigating the impact of disruptions.

6.5.10 CONCLUSION

The mathematical exploration of various network topologies and their spectral properties provides a comprehensive understanding of their impact on supply chain connectivity and robustness. By leveraging these mathematical insights, logisticians and engineers can design and optimize supply chain networks that are not only efficient and cost-effective but also resilient and adaptable to dynamic operational conditions.

The subsequent sections will delve deeper into specific network configurations, optimization techniques, and real-world applications, further bridging the gap between theoretical network science and practical logistics management.

6.6 NETWORK CONFIGURATIONS AND THEIR IMPLICATIONS ON LOGISTICS

Continuing the discourse on network topologies, this section provides a detailed mathematical analysis of various network configurations—random, small-world, scale-free, and hierarchical—and their specific implications for logistics and supply chain management. By dissecting each topology's structural and spectral properties, we can ascertain their suitability for different logistical applications and identify strategies for optimizing supply chain performance.

6.6.1 RANDOM NETWORKS IN LOGISTICS

Random networks serve as a foundational model in network science, offering insights into the behavior of systems with no inherent structure. In logistics, random networks can model scenarios where connections between entities are formed without strategic planning or constraints.

6.6.1.1 Erdős-Rényi Model $G(n,p)$

The Erdős-Rényi model $G(n,p)$ generates random graphs by connecting each pair of n nodes with independent probability p. The expected degree λ of each node is $\lambda = p(n-1)$.

The degree distribution $P(k)$ in random networks follows a binomial distribution, which approximates a Poisson distribution for large n and small p:

$$P(k) = \binom{n-1}{k} p^k (1-p)^{n-1-k} \approx \frac{\lambda^k e^{-\lambda}}{k!}.$$

The eigenvalues of the Laplacian matrix L in random networks tend to cluster around the mean degree λ. The algebraic connectivity λ_2 increases with λ, enhancing network robustness.

1. **Cost Inefficiency:** The lack of strategically placed hubs leads to inefficient routing and higher transportation costs.
2. **Low Robustness:** Random networks exhibit lower λ_2, making them more vulnerable to random failures.
3. **Limited Scalability:** As n increases, maintaining λ_2 requires proportionally increasing p, which may not be feasible.

6.6.2 SMALL-WORLD NETWORKS IN LOGISTICS

Small-world networks, characterized by high clustering and short path lengths, are particularly suited for logistics networks that require both localized efficiency and global connectivity.

6.6.2.1 Watts-Strogatz Model $G(n,k,p)$

The Watts-Strogatz model generates small-world networks by starting with a regular ring lattice with n nodes and each node connected to k nearest neighbors. Edges are then randomly rewired with probability p, introducing shortcuts that reduce average path lengths.

The clustering coefficient C in small-world networks remains high, similar to regular lattices:

$$C \approx 3\frac{n}{k(k-1)}.$$

This high clustering facilitates efficient localized operations and information sharing within the supply chain.

The average path length L scales logarithmically with n, even for small p, significantly reducing the number of steps required for traversal:

$$L \sim \log n.$$

This property ensures swift dissemination of information and resources across the network.

Small-world networks exhibit enhanced algebraic connectivity λ_2 due to the introduction of shortcuts, which act as critical bridges enhancing overall network resilience.

The implications for logistics are:

1. **Balanced Efficiency:** Small-world topologies strike a balance between local operational efficiency and global connectivity, optimizing resource distribution.
2. **Moderate Robustness:** The presence of shortcuts enhances robustness against random failures, although targeted attacks on shortcut edges can still disrupt the network.
3. **Flexibility and Scalability:** The modular structure allows for scalability, accommodating growth without compromising structural integrity.

6.6.3 SCALE-FREE NETWORKS IN LOGISTICS

Scale-free networks, characterized by a power-law degree distribution, are prevalent in many real-world supply chains where certain entities serve as central hubs.

6.6.3.1 Barabasi-Albert Model $G(n,m)$

The Barabási-Albert model generates scale-free networks through preferential attachment, where new nodes preferentially connect to existing nodes with higher degrees.

Scale-free networks follow a power-law degree distribution:

$$P(k) \sim k^{-\gamma},$$

with $\gamma \approx 3$, leading to the emergence of hubs—nodes with significantly higher degrees compared to others.

The presence of hubs in scale-free networks increases λ_2, enhancing algebraic connectivity and overall network robustness. However, the dependence on hubs introduces vulnerabilities, as the removal of a hub can drastically reduce λ_2, fragmenting the network.

The implications for logistics are:

1. **Operational Efficiency:** Hubs facilitate bulk operations and centralized resource allocation, reducing operational costs and enhancing efficiency.
2. **Centralized Control:** The network's centralization allows for streamlined decision-making and coordinated supply chain activities.
3. **Vulnerability to Hub Failures:** The reliance on hubs necessitates robust protection strategies to prevent supply chain disruptions due to hub failures.

6.6.4 HIERARCHICAL NETWORKS IN LOGISTICS

Hierarchical networks mirror the multi-tiered structure of supply chains, embodying a clear organizational hierarchy from suppliers to end consumers.

6.6.4.1 Hierarchical Graph Model

Hierarchical networks are structured into multiple layers, each representing different functional levels within the supply chain. Mathematically, a hierarchical graph can be modeled as $G = (V, E)$ where:

1. V is partitioned into layers $V_1, V_2, \ldots, V_L$.
2. E predominantly connects nodes between adjacent layers, preserving the hierarchical flow.

Hierarchical networks exhibit varying spectral properties based on inter-layer connectivity and redundancy. Enhanced connectivity between layers increases λ_2, while redundancy within layers improves robustness.

The implications for logistics are:

1. **Structured Operations:** The hierarchical structure promotes organized resource distribution and streamlined operational processes.
2. **Scalability:** Hierarchical networks can scale efficiently by adding additional layers or expanding existing ones without compromising network stability.
3. **Potential Bottlenecks:** Centralized tiers within the hierarchy can become bottlenecks, requiring strategic management to prevent operational delays.

6.6.5 COMPARATIVE SPECTRAL ANALYSIS

A comparative analysis of the spectral properties across different network topologies elucidates their respective strengths and vulnerabilities in logistics applications.

6.6.5.1 Algebraic Connectivity Comparison

$$\lambda_2^{\text{random}} < \lambda_2^{\text{small-world}} < \lambda_2^{\text{scale-free}} \approx \lambda_2^{\text{hierarchical}}$$

Scale-free and hierarchical networks typically exhibit higher λ_2 compared to small-world and random networks, indicating superior connectivity and robustness.

6.6.5.2 Spectral Gap Analysis

The spectral gap $\Delta = \lambda_2 - \lambda_1$ reflects the network's robustness:

$$\Delta^{\text{scale-free}} > \Delta^{\text{small-world}} > \Delta^{\text{random}} > \Delta^{\text{hierarchical}}$$

A larger spectral gap implies greater resilience against perturbations and failures.

6.6.6 OPTIMIZATION STRATEGIES BASED ON SPECTRAL PROPERTIES

Leveraging spectral properties to inform optimization strategies can significantly enhance supply chain performance.

6.6.6.1 Maximizing Algebraic Connectivity

Formulate the optimization problem to maximize λ_2:

$$\max_G \quad \lambda_2(G),$$

$$\text{subject to:} \quad |E| \leq E_{\max},$$

where $|E|$ is the number of edges and $E_{\max}$ the maximum allowed edges. This ensures the network remains highly connected while adhering to cost constraints.

6.6.6.2 Minimizing Spectral Gap

Alternatively, minimize the spectral gap to optimize synchronization and information flow:

$$\min_G \quad \Delta(G),$$

$$\text{subject to:} \quad |E| \geq E_{\min},$$

where $E_{\min}$ ensures sufficient connectivity.

Implement algorithms such as gradient ascent for maximizing λ_2, or eigenvalue optimization techniques to iteratively adjust the network topology for optimal spectral properties.

6.6.7 IMPLICATIONS FOR SUSTAINABLE MANUFACTURING

Optimizing network topologies through spectral analysis not only enhances connectivity and robustness but also contributes to sustainable manufacturing practices by enabling efficient resource allocation and minimizing environmental impact.

6.6.7.1 Energy-Efficient Network Designs

Mathematically, design networks that minimize energy consumption while maintaining high λ_2:

$$\min_{G} \sum_{e \in E} w_e f_e,$$

$$\text{subject to:} \quad \lambda_2(G) \geq \lambda_{\min},$$

where w_e represents the energy cost of edge e and f_e the flow through edge e.

Conduct a trade-off analysis between energy consumption and network robustness to identify optimal configurations that balance sustainability with operational efficiency.

6.6.8 CONCLUSION

This section has provided a comprehensive mathematical analysis of various network topologies and their implications for logistics and supply chain management. By examining the structural and spectral properties of random, small-world, scale-free, and hierarchical networks, we have elucidated their respective strengths and vulnerabilities in logistical applications. The comparative spectral analysis underscores the significance of algebraic connectivity and spectral gaps in assessing network robustness and efficiency.

Mathematically optimizing network topologies based on spectral properties offers a pathway to designing supply chains that are not only efficient and cost-effective but also resilient and sustainable. As we progress through this chapter, we will further explore optimization strategies, delve into real-world case studies, and examine the integration of advanced network metrics to provide a holistic understanding of network science applications in logistics.

6.7 SPECTRAL PROPERTIES OF GRAPHS AND THEIR ROLE IN SUPPLY CHAIN ROBUSTNESS

The spectral properties of graphs, particularly those derived from the Laplacian matrix, are instrumental in evaluating and enhancing the robustness of supply chain networks. This section delves into the mathematical intricacies of spectral graph theory and elucidates how these properties inform the design of resilient logistics systems.

6.7.1 GRAPH LAPLACIAN AND ITS EIGENVALUES

The Laplacian matrix L encapsulates the connectivity of a graph and serves as a fundamental tool in spectral graph theory. Defined as:

$$L = D - A,$$

where D is the degree matrix and A the adjacency matrix, L is a symmetric, positive semi-definite matrix with eigenvalues $0 = \lambda_1 \leq \lambda_2 \leq \cdots \leq \lambda_n$.

6.7.1.1 Algebraic Connectivity (λ_2)

The second smallest eigenvalue λ_2, or algebraic connectivity, is a critical indicator of the graph's overall connectivity. Mathematically:

$$\lambda_2 = \min_{\mathbf{x} \perp \mathbf{1}} \frac{\mathbf{x}^\top L \mathbf{x}}{\mathbf{x}^\top \mathbf{x}}.$$

A higher λ_2 signifies a more robust network, resistant to fragmentation and efficient in resource distribution.

6.7.1.2 Spectral Gap (Δ)

The spectral gap, defined as $\Delta = \lambda_2 - \lambda_1$, where $\lambda_1 = 0$ for connected graphs, serves as a measure of the network's resilience:

$$\Delta = \lambda_2.$$

A larger spectral gap implies greater robustness against node and edge removals, enhancing the supply chain's ability to withstand disruptions.

6.7.2 FIEDLER VECTOR AND NETWORK PARTITIONING

The Fiedler vector $\mathbf{v}_2$, corresponding to λ_2, provides invaluable insights into the network's community structure and potential weak points.

Graph Partitioning Using Fiedler Vector

Partition the network into two subgraphs by leveraging the signs or magnitudes of the components of $\mathbf{v}_2$:

$$\text{Partition: } V_1 = \{v_i \in V \mid (\mathbf{v}_2)_i \geq 0\},$$

$$V_2 = \{v_i \in V \mid (\mathbf{v}_2)_i < 0\}.$$

This partitioning strategy identifies clusters within the supply chain that are densely interconnected internally but sparsely connected externally, highlighting potential areas for redundancy enhancement.

6.7.2.1 Implications for Supply Chain Design

By analyzing the Fiedler vector, logisticians can:

1. **Identify Critical Nodes:** Nodes with high components in $\mathbf{v}_2$ are pivotal for maintaining network connectivity.
2. **Optimize Network Structure:** Reinforce connections between clusters to prevent fragmentation and enhance overall robustness.
3. **Facilitate Targeted Investments:** Allocate resources strategically to strengthen weak points identified through spectral analysis.

6.7.3 SPECTRAL GRAPH THEORY APPLICATIONS IN LOGISTICS

Spectral graph theory provides a suite of mathematical tools that are directly applicable to logistics network optimization.

6.7.3.1 Network Robustness Assessment

Evaluate the robustness of supply chain networks by computing λ_2 and analyzing how it changes under various perturbations, such as node or edge removals:

$$\lambda_2(G - \{v_i\}) \geq \lambda_2(G) - \delta,$$

where δ represents the impact of removing node v_i.

Define resilience metrics based on spectral properties to quantify the network's ability to recover from disruptions:

$$R(G) = \frac{\lambda_2(G')}{\lambda_2(G)},$$

where G' is the perturbed network after disruptions.

6.7.3.2 Flow Optimization and Synchronization

Utilize spectral properties to optimize the flow of goods and information within the network, ensuring synchronized operations and minimizing delays.

Formulate flow equations that incorporate spectral metrics to enhance synchronization:

$$\mathbf{f} = -L^{\dagger}\mathbf{d},$$

where $L^{\dagger}$ is the pseudo-inverse of the Laplacian matrix and $\mathbf{d}$ is the demand vector.

Measure the degree of synchronization in the network by evaluating the alignment of flow vectors with the Fiedler vector:

$$S = \frac{\mathbf{f}^{\top}\mathbf{v}_2}{\|\mathbf{f}\|\|\mathbf{v}_2\|}.$$

Higher S values indicate better synchronization, leading to more efficient supply chain operations.

6.7.3.3 Community Detection and Modular Network Design

Employ spectral clustering techniques to detect communities within the logistics network, facilitating modular design and localized optimization.

Execute the following steps for spectral clustering:

1. Compute the Laplacian matrix L of the network.
2. Determine the Fiedler vector $\mathbf{v}_2$.

3. Partition the network based on the median or sign of $\mathbf{v}_2$.
4. Iteratively refine partitions to maximize intra-community connectivity and minimize inter-community connections.

The implications for logistics are:

1. **Localized Optimization:** Enable targeted optimization within communities, enhancing efficiency without disrupting the entire network.
2. **Scalability:** Facilitate scalable supply chain designs by compartmentalizing the network into manageable modules.
3. **Enhanced Resilience:** Improve overall network resilience by ensuring that each community operates robustly, independent of others.

6.7.4 CASE STUDY: ENHANCING SUPPLY CHAIN ROBUSTNESS THROUGH SPECTRAL ANALYSIS

To illustrate the practical application of spectral graph theory in logistics, consider a hypothetical supply chain network consisting of multiple suppliers, manufacturers, distribution centers, and retailers.

6.7.4.1 Network Configuration

- **Nodes:** 50 suppliers, 20 manufacturers, 10 distribution centers, 100 retailers.
- **Edges:** Represent transportation routes between nodes, weighted by transportation cost and capacity.
- **Initial Topology:** Modeled as a small-world network to balance efficiency and robustness.

6.7.4.2 Spectral Analysis

1. **Compute Laplacian Matrix:** Derive L from the adjacency matrix A and degree matrix D.
2. **Determine Eigenvalues and Fiedler Vector:** Calculate λ_2 and $\mathbf{v}_2$.
3. **Identify Critical Nodes:** Analyze $\mathbf{v}_2$ to detect nodes with high centrality and potential vulnerability.

Findings:

1. **Algebraic Connectivity:** $\lambda_2 = 0.45$, indicating a moderately robust network.
2. **Critical Nodes Identified:** Three distribution centers emerge as critical hubs with high betweenness centrality.
3. **Vulnerability Assessment:** Simulations show that the removal of any critical node reduces λ_2 below 0.3, substantially compromising network robustness.

Some optimization recommendations are:

1. **Redundancy Enhancement:** Introduce additional transportation routes connecting critical distribution centers to increase λ_2.
2. **Hub Protection:** Implement protective measures for critical nodes, such as backup facilities or alternative suppliers, to prevent network fragmentation.
3. **Modular Design:** Restructure the network into clusters based on the Fiedler vector to facilitate localized resilience and targeted optimization.

6.7.5 CONCLUSION

The spectral properties of graphs offer a mathematically robust framework for evaluating and enhancing the robustness of supply chain networks. By analyzing eigenvalues and eigenvectors of the Laplacian matrix, logisticians can identify critical nodes, optimize network topologies, and implement strategies that bolster network resilience. This mathematical approach not only aids in preemptively mitigating vulnerabilities but also facilitates the design of supply chains that are efficient, scalable, and adaptable to dynamic operational conditions. The subsequent sections will further explore advanced network configurations, optimization techniques, and their direct applications in real-world logistics scenarios, underscoring the integral role of network science in modern supply chain management.

6.8 GRAPH CONFIGURATIONS AND THEIR IMPACT ON SUPPLY CHAIN EFFICIENCY

Continuing our exploration of network topologies, this section delves deeper into the specific configurations of graphs and their mathematical implications for supply chain efficiency and robustness. By understanding the nuanced properties of different network structures, logisticians can strategically design and optimize supply chains to meet operational demands and withstand potential disruptions.

6.8.1 RANDOM NETWORKS REVISITED: LIMITATIONS AND APPLICATIONS

While random networks provide a foundational understanding of network behavior, their application in logistics is often limited due to inherent inefficiencies and vulnerabilities. However, certain logistical scenarios, such as ad-hoc supply chains formed during emergency responses, can benefit from the simplicity and flexibility of random network configurations.

6.8.1.1 Mathematical Analysis of Random Network Vulnerabilities

Consider a random network $G(n, p)$ with n nodes and edge probability p. The probability that the network is connected is:

$$P(\text{connected}) \approx 1 - ne^{-\lambda},$$

for $\lambda = p(n-1) \gg \ln n$. However, as p decreases, the network becomes increasingly fragmented, with clusters forming based on probabilistic edge connections.

The percolation threshold p_c for connectivity in a random network is approximately:

$$p_c \approx \frac{\ln n}{n}.$$

Below this threshold, the network almost surely becomes disconnected, highlighting the fragility of random supply chains.

6.8.1.2 Implications for Logistics Design

The inherent randomness and lack of strategic connectivity in random networks result in high transportation costs, inefficient resource allocation, and low resilience. Therefore, while random networks offer theoretical insights, practical logistics applications necessitate more structured and optimized network configurations.

6.8.2 SMALL-WORLD NETWORKS: ENHANCING SUPPLY CHAIN AGILITY

Small-world networks, with their unique balance of local clustering and global shortcuts, offer significant advantages for supply chain agility and efficiency.

6.8.2.1 Mathematical Properties Facilitating Agility

In small-world networks $G(n,k,p)$, the introduction of random shortcuts with probability p exponentially decreases the average path length L, enhancing the speed of information and resource flow:

$$L \sim \frac{\ln n}{\ln \langle k \rangle},$$

where $\langle k \rangle$ is the average degree. Simultaneously, the clustering coefficient C remains high, promoting efficient local operations.

6.8.2.2 Spectral Implications

The enhanced algebraic connectivity λ_2 in small-world networks correlates with improved robustness against random failures:

$$\lambda_2^{\text{small-world}} \gg \lambda_2^{\text{random}}.$$

This increased λ_2 ensures that the network remains largely connected even when random nodes or edges are removed, thereby enhancing supply chain resilience.

6.8.2.3 Applications in Logistics

Small-world network configurations are particularly suited for:

1. **Distributed Manufacturing:** Facilitating rapid coordination and resource sharing across geographically dispersed manufacturing units.

2. **Real-Time Information Sharing:** Enabling swift dissemination of information and demand signals across the supply chain, reducing lead times and enhancing responsiveness.
3. **Decentralized Decision-Making:** Supporting localized decision-making processes while maintaining overall network efficiency through global shortcuts.

6.8.3 SCALE-FREE NETWORKS: CENTRALIZATION VS. ROBUSTNESS

Scale-free networks present a paradox in supply chain design: the efficiency gains from centralized hubs come at the cost of increased vulnerability.

6.8.3.1 Mathematical Trade-Offs

Scale-free networks optimize resource distribution through highly connected hubs, leading to lower average path lengths and higher λ_2:

$$\lambda_2^{\text{scale-free}} > \lambda_2^{\text{small-world}} > \lambda_2^{\text{random}}.$$

However, the heavy-tailed degree distribution introduces critical vulnerabilities. The probability of random node removal $P(v_i)$ is inversely proportional to node degree k_i:

$$P(v_i) \sim k_i^{-\gamma}.$$

This means that high-degree hubs are rare but crucial for maintaining network connectivity.

6.8.3.2 Spectral Vulnerabilities

The removal of a hub node in a scale-free network causes a significant drop in λ_2, potentially fragmenting the network into isolated clusters:

$$\lambda_2(G - \{v_i\}) \approx \lambda_2(G) - \delta,$$

where δ is substantial for high-degree v_i.

6.8.3.3 Applications and Mitigation Strategies

Scale-free network configurations are advantageous for:

1. **Centralized Distribution Centers:** Enhancing bulk operations and reducing overall transportation costs.
2. **Strategic Resource Allocation:** Concentrating resources in hubs to optimize supply chain performance.

To mitigate vulnerabilities:

1. **Hub Redundancy:** Introduce backup hubs to ensure continuity in case of hub failures.
2. **Decentralization:** Gradually decentralize operations to distribute load and reduce dependence on single points of failure.

6.8.4 HIERARCHICAL NETWORKS: STRUCTURING MULTI-TIERED SUPPLY CHAINS

Hierarchical networks model the multi-tiered nature of complex supply chains, providing a structured approach to resource distribution and operational management.

6.8.4.1 Mathematical Structuring of Hierarchical Networks

Hierarchical networks are organized into layers $V_1, V_2, \ldots, V_L$, each representing different functional levels within the supply chain:

1. V_1: Raw material suppliers.
2. V_2: Manufacturers.
3. V_3: Distributors.
4. V_4: Retailers.

Edges E predominantly connect nodes between adjacent layers, maintaining a clear hierarchical flow of goods and information.

The modular nature of hierarchical networks allows for localized spectral analysis within each layer, facilitating targeted optimization:

$$L = \begin{bmatrix} L_1 & 0 & \cdots & 0 \\ 0 & L_2 & \cdots & 0 \\ \vdots & \vdots & \ddots & \vdots \\ 0 & 0 & \cdots & L_L \end{bmatrix},$$

where L_i is the Laplacian matrix for layer V_i.

6.8.4.2 Implications for Supply Chain Efficiency

1. **Structured Resource Allocation:** Hierarchical networks facilitate organized distribution and production processes, enhancing efficiency.
2. **Scalability:** Adding new nodes or layers can be managed without disrupting the overall network structure.
3. **Potential for Bottlenecks:** Centralized layers may become points of congestion, requiring mathematical optimization to distribute load evenly.

6.8.5 GRAPH CONFIGURATIONS AND DEGREE DISTRIBUTIONS

The degree distribution $P(k)$ significantly influences the network's spectral properties and logistical performance.

6.8.5.1 Scale-Free vs. Small-World vs. Random Degree Distributions

1. **Scale-Free:** Heavy-tailed distribution with hubs, enhancing λ_2 but introducing vulnerability.
2. **Small-World:** Moderately heterogeneous distribution with clustering and shortcuts, balancing efficiency and robustness.

3. **Random:** Homogeneous distribution, low clustering, high vulnerability, and inefficiency.

Different degree distributions result in varying spectral gaps and connectivity measures, directly impacting supply chain resilience and operational efficiency.

6.8.5.2 Power-Law vs. Poisson vs. Exponential Degree Distributions

1. **Power-Law:** $P(k) \sim k^{-\gamma}$, leading to hub dominance and high algebraic connectivity.
2. **Poisson:** $P(k) = \frac{\lambda^k e^{-\lambda}}{k!}$, indicating uniform connectivity and low λ_2.
3. **Exponential:** $P(k) \sim e^{-\beta k}$, providing a balance between uniformity and heterogeneity.

The impact on logistics are:

Power-law distributions facilitate efficient centralized operations but require robust strategies to protect hubs. Poisson distributions, while simple, lack the structural advantages necessary for efficient supply chain operations. Exponential distributions offer a middle ground, providing some degree of heterogeneity without extreme centralization.

6.8.6 ADVANCED MATHEMATICAL TECHNIQUES FOR NETWORK OPTIMIZATION

To harness the full potential of network science in logistics, advanced mathematical techniques are employed to optimize network structures based on spectral properties and degree distributions.

6.8.6.1 Eigenvalue Optimization

Optimize the eigenvalues of the Laplacian matrix to enhance network robustness and efficiency:

$$\max_{G} \quad \lambda_2(G),$$

$$\text{subject to:} \quad |E| \leq E_{\max},$$

where G is the graph configuration, and $E_{\max}$ is the maximum number of edges.

Employ numerical optimization algorithms to iteratively adjust network topology, maximizing λ_2 while adhering to logistical constraints such as budget and operational capacity.

6.8.6.2 Combinatorial Optimization for Degree Distribution

Manipulate the degree distribution to achieve desired network properties:

$$\min_{P(k)} \quad f(\lambda_2, \Delta),$$

$$\text{subject to:} \quad \sum_k P(k) = 1, \quad \sum_k kP(k) = \langle k \rangle,$$

where f is a function that encapsulates the trade-off between algebraic connectivity λ_2 and spectral gap Δ.

Tailoring the degree distribution allows for strategic placement of hubs and the creation of efficient pathways, optimizing both connectivity and robustness.

6.9 CONCLUSION

The integration of network science into logistics and supply chain management represents a paradigm shift toward more analytical and mathematically grounded approaches. By modeling supply chains as complex networks, leveraging spectral graph theory, and employing optimization techniques, logisticians can gain profound insights into the structural and dynamic aspects of their operations. This chapter has laid the mathematical foundations necessary for understanding and applying network science principles to logistics, highlighting the critical role of network topologies, spectral properties, and advanced metrics in enhancing supply chain efficiency and robustness.

Through the examination of random, small-world, scale-free, and hierarchical networks, we have elucidated the strengths and limitations of each topology in the context of logistics. The spectral analysis has demonstrated how eigenvalues and eigenvectors can be harnessed to evaluate and optimize network connectivity, identify critical nodes, and enhance resilience against disruptions. Furthermore, the exploration of advanced network metrics and optimization strategies has provided a roadmap for designing supply chains that are not only cost-effective and efficient but also robust and adaptable to the ever-evolving demands of the global market.

As supply chains continue to grow in complexity and globalization, the mathematical rigor and analytical depth provided by network science will be indispensable in navigating logistical challenges and achieving operational excellence. The subsequent chapters will build upon this foundation, offering detailed examinations of specific network configurations, optimization techniques, and real-world applications that underscore the transformative potential of network science in logistics.

Students equipped with the knowledge and mathematical expertise presented in this chapter will be well-prepared to innovate and drive advancements in supply chain management, contributing to the development of resilient, efficient, and sustainable logistics systems in an increasingly interconnected world.

Section III

Case Studies, Real-World Applications, Emerging Trends, and Advanced Topics

7 Case Studies and Real-World Applications

7.1 INTRODUCTION

The goal of this chapter is the analysis and application of theoretical concepts to real-world logistic problems such as those found in different case studies from Advanced Manufacturing, Rapid Prototyping, and Healthcare industries. Logistics optimization is essential for modern supply chain networks. This chapter explores how **network science** principles enhance logistics operations through **real-world case studies**, emphasizing **efficiency, resilience, and sustainability**.

7.2 CASE STUDIES AND REAL-WORLD APPLICATIONS

The following case studies illustrate practical applications of network science and automatic control in logistics.

7.2.0.1 Case Study: Global Automotive Supply Chain

Tesla is an American electric-car company based in Palo Alto, California, while Tesla, Inc. is an American multinational clean energy and electric vehicle (EV) company headquartered in Austin, Texas. It is a leader in EV technology, producing electric cars (both passenger and commercial trucks), battery energy storage systems from residential to grid-scale, solar (photovoltaic) panels, and solar-integrated roofing products, among other related technologies and services. Tesla is the world's most valuable company, retaining its status as the world's number one automaker with a market capitalization of more than USD 550 billion. Tesla, in 2021, became a world leader in the sale of battery electric vehicles (BEVs), having 21% market share in the BEV market and 14% in the total plug-in electric vehicle (PEV) market, including plug-in hybrid electric vehicles (PHEVs). With its subsidiary Tesla Energy, the company is now a forefront renewable energy innovator, standing among the largest developers and installers of photovoltaic energy systems in the United States. Tesla Energy is also a leading global supplier of battery energy storage systems (BESS), having 3.99 gigawatt-hours (GWh) installed in 2021.

Supply chain management (SCM) is particularly crucial within the vehicle manufacturing industry, most notably the transforming new energy vehicle (NEV) sector. The intricacy of electric vehicle production supply chains has been a crucial challenge, where advanced automation, logistics optimization, and strong procurement measures must be adopted in order to make it efficient and sustainable in worldwide operations (see Figure 7.1).

DOI: 10.1201/9781003627906-7

Figure 7.1 Integrated view from Energy generation to mobility application.

The business was formed under the name Tesla Motors, Inc., on July 1, 2003, by Martin Eberhard and Marc Tarpenning. Eberhard and Tarpenning assumed the positions of CEO and CFO, respectively. Eberhard explained that he wanted to have a car manufacturer that is a technology company, where the central technologies are the battery, the computer software, and the special motor. Ian Wright was Tesla's third hire, showing up a few months later. In February 2004, the company raised USD 7.5 million in Series A funding, USD 6.5 million of which was provided by Elon Musk, who had received USD 100 million from the sale of his interest in PayPal two years earlier. Musk took the position of chairman of the board of directors and the most astronomically gigantic shareholder of Tesla. J. B. Straubel came on board at Tesla in May 2004 as a chief technical officer [26]. Musk took the position of chairman of the board of directors and the most astronomically gigantic shareholder of Tesla. J. B. Straubel came on board at Tesla in May 2004 as a chief technical officer. Beneficiary of excellent timing, Tesla effectively was the only company building battery-electric cars when fuel prices spiked in the summer of 2008; as a result, no news story on alternative fuels was complete without mentioning Tesla.

The company has expanded from a maker of electric cars to one that also makes energy-related products like solar roof panels and energy storage systems. Current vehicles include the Model 3 compact sedan, Model S mid-size luxury sedan, Model X luxury SUV, and Model Y compact SUV. The automaker previously built the Tesla Roadster, a two-seat convertible that cost more than USD 100,000 when new.

Tesla Motors has become known for manufacturing well-engineered and innovative electric conveyances. To assemble its cars, Tesla buys parts from a range of Tesla suppliers. Some of its suppliers sell generic car parts; others sell electric conveyance parts. Others supply steel, industrial products, and minerals. The transition to EVs is moving at a very fast pace, and EV suppliers have room to grow. As the EV industry expands, Tesla suppliers are potential investment targets. Tesla's production centers are in North America, the EU, and China. Similar to most automotive companies, Tesla's supply chain strategy is founded on purchasing components and systems from a single supplier. Where there are available multiple sources for particular critical components, Tesla strives to qualify multiple suppliers in order to reduce production risks from supply disruptions [30].

Tesla's entire financial and operational system is entirely internally funded and centralized using the ERP software built internally under the label Warp. Warp enables real-time collaborative and synchronized operation for a seamless workflow in Tesla production and staff functions.

Tesla's conversion to a company-owned store and service center model is tactically driven by its quest to maximize supply chain efficiency and provide high-priority customer service. By maintaining direct ownership of its sales and service infrastructure, Tesla aims to optimize logistics coordination and provide a seamless ownership experience. Up to early 2012, Tesla had 10 stores, 1 gallery, and 9 service centers in the United States. By the end of the year, the company projected growth to 19 stores, 3 galleries, and 26 service centers. Unexpectedly, Tesla's service centers outnumber its stores and galleries together, indicating how much the company values after-sales support and ease of maintenance. Moreover, Tesla has established service centers in cities where there are no physical retail branches, which enables wider coverage of areas where customers can receive support.

Tesla's Shanghai Gigafactory has played a crucial role in the company's international supply chain strategy, particularly in export operations. During the first half of 2022, Model Y and Model 3 vehicle exports from Gigafactory Shanghai accounted for 50% of China's total electric vehicle exports, indicating the dominance of Tesla in overseas EV distribution. Tesla has a vertically integrated supply chain, as opposed to the traditional automobile manufacturers, that provides greater control over network design, supplier relationships, and strategic logistics activities. As opposed to the traditional supply chains, Tesla's model illustrates the implications of vertical integration on operational efficiency, cost savings, and responsiveness to the market. Tesla's supply chain model provides valuable business strategy and logistics innovation lessons. Through establishing robust partnerships with vendors and having end-to-end supply chain management, Tesla enhances the supply chain's resilience and aligns its logistics strategy with overall corporate performance [23]. The model serves as a reference for organizations that wish to see how cooperative vendor relationships contribute to strategic supply chain objectives and staying ahead in the marketplace. The Tesla automobile supply chain network is divided into three layers: Tesla enterprises, direct suppliers of automobile parts, and indirect suppliers.

Analyze a global automotive supply chain network structured as a scale-free network. Utilize spectral graph analysis to identify critical suppliers and optimize transportation routes to enhance λ_2, thereby improving overall network robustness.

Relevant findings are:

1. **Critical Supplier Identification:** Spectral analysis revealed key suppliers acting as hubs, crucial for maintaining supply chain continuity.
2. **Route Optimization:** Enhancing connections between critical hubs and secondary suppliers significantly increased λ_2, improving resilience against supplier disruptions.

The degree of the node enterprise is the connection relationship with other enterprises in the Tesla automobile supply chain network, and the larger the degree value, the more enterprises with which it generates supply relationships. In the Tesla automobile supply chain network, Tesla enterprises has the largest degree value and is in the core position, and all other enterprises have a supply relationship with Tesla enterprises directly or indirectly. In addition, according to the new energy vehicle parts and components system, it is divided into a power train system, central control system, electric drive system, charging system, chassis and body, interior and exterior decoration, and other modules. The density of this network is 0.019, and the connection between node enterprises is not particularly close, but the average degree is high, 2.336, and each node enterprise in the Tesla car supply chain network has business relations with an average of 2.633 node enterprises [24].

The K-shell identification method [8] based on domain improvement, which first stratifies the nodes by the K-shell method to determine the influence of the global structure of the nodes can be used. Then measures the importance of the nodes according to the influence of the node's first reachable and second reachable points, and defines the node's importance index as follows:

$$K_s(i) = EC1(i) * \delta + EC2(i) * \delta^2 \tag{7.1}$$

where $EC1(i)$ is the average eigenvector value of the first reachable point of the node, $EC2(i)$ is the average eigenvector value of the second reachable point of the node, and $\delta \in (0,1)$ is the scaling factor. For a node in a network, both neighboring nodes and distant nodes have an influence on that node, although each has a different degree of influence.

Within the automotive supply chain network, the level of connectivity can be used to illustrate the connectivity between node businesses: the higher the number of node businesses, the larger their network size, and the higher the level of connectivity among node businesses in the automotive supply chain network. To describe the degree of direct change in the connectivity of the automotive supply chain network after being attacked, a node attack sequence "A" is fixed beforehand, and when nodes are attacked in network one by one according to the node order in "A," the size of the maximum connectivity component of network will decrease as the attack goes on. Individual node failures in a supply chain network are subject to load propagation and load redistribution. Load propagation leads to new node failures due to being overloaded, load redistribution, and node-node interactions, hence the cascading failure of the overall supply chain network. The three elements of cascade failure model are node initial load, node capacity, and load redistribution strategy. A key engineering contribution predicated on mathematical basis is the analysis of supply chain disruptions. Disruptions can occur due to numerous reasons like cyberattacks, natural calamities, failures in suppliers, and logistics bottlenecks. One of the bigger supply chain management challenges is to understand how a sequence of attacks—strategic or random—can collapse or destabilize an entire logistics system.

To see how disruptions propagate, we can think of a supply chain as a complex network where suppliers, manufacturers, warehouses, and distribution centers are the nodes connected to each other and transportation routes are the edges. When a critical node or link fails, its failure can have cascading effects that halt the flow of goods, delay production schedules, and incur operational expenses. The effect of these disruptions can be measured using algebraic connectivity, which describes how well-connected a network remains after each failure. Mathematically, the Laplacian matrix of a graph provides an eigenvalue known as the Fiedler value, which decreases as nodes are removed. A diminishing Fiedler value signals an increasingly fragmented network, leading to reduced redundancy and alternative routing options. As a network loses its connectivity, supply chain operations become less resilient, requiring longer transportation routes and increasing costs.

A cascading failure model helps in understanding how an initial disruption forces the redistribution of logistical loads, overloading alternative routes and causing secondary failures. If a critical node in the supply chain is removed, neighboring nodes experience an increase in demand pressure. If these neighboring nodes exceed their capacity, they also fail, triggering further breakdowns. This behavior can be modeled using load redistribution equations, where the failure of a node causes its demand to be reallocated among surviving nodes based on their remaining capacity. The rate at which failures propagate depends on the network topology and the flexibility of its components. Analysis of the evolution of the network can help in the implementation of reactive measures, which is one of the key contributions of having such engineering-based perspective of the logistics.

7.2.0.2 Case Study: E-Commerce Distribution Network

Amazon operates one of the most advanced and automated e-commerce distribution networks in the world. This network integrates robotics, artificial intelligence, and real-time data analytics to optimize supply chain efficiency and enhance customer satisfaction. The system consists of multiple interlinked components, each playing a crucial role in ensuring seamless order fulfillment. The primary nodes in Amazon's logistics network include fulfillment centers, sortation centers, last-mile delivery stations, and return processing hubs. At fulfillment centers, goods are stored, retrieved, and packed before being dispatched to sortation centers, where they are categorized based on their final destination. Last-mile delivery stations handle local distribution, ensuring timely deliveries to customers, while return processing hubs manage product returns and re-integration into the supply chain [12].

A significant aspect of Amazon's logistics efficiency is its automation-driven approach. Within fulfillment centers, Automated Storage and Retrieval Systems (AS/RS) are implemented using Kiva robots, which autonomously transport shelves to human pickers, reducing the need for manual warehouse navigation. The integration of AI-powered demand forecasting allows Amazon to predict customer orders with high accuracy, optimizing inventory placement across various fulfillment centers to minimize shipping times. Furthermore, dynamic route optimization enhances last-mile delivery by analyzing traffic patterns, weather conditions, and delivery urgencies, ensuring that packages reach customers in the shortest time possible. Robotic Process Automation (RPA) streamlines repetitive tasks such as sorting and packing, allowing for higher efficiency and lower operational costs. In addition, the use of IoT and RFID tracking provides real-time visibility into package locations, reducing instances of lost shipments and improving overall transparency [21].

The distribution network can be mathematically characterized using graph theory, where Amazon's logistics infrastructure is represented as a directed weighted graph $G = (V, E)$. In this model, V represents the nodes, including fulfillment centers, sortation hubs, delivery stations, and customers, while E represents the transportation routes connecting these nodes. Each edge e_{ij} carries a weight w_{ij} that accounts for delivery time, transportation costs, and congestion levels between nodes $i, j \in E$. The primary objective of Amazon's logistics optimization is to minimize the total delivery cost, which can be expressed as a mathematical function:

$$min(\sum_{(i,j)\in E} x_{ij} \cdot w_{ij}) \tag{7.2}$$

where x_{ij} is a binary decision variable that indicates whether a package is routed through edge (i, j).

$$x_{ij} = \begin{cases} 1, & \text{if } path(i,j) \text{ is used} \\ 0, & \text{otherwise} \end{cases} \tag{7.3}$$

The fulfillment centers, being critical nodes in the system, can be modeled using M/M/c queuing theory, where incoming orders are treated as a Poisson process with an arrival rate λ, and service times follow an exponential distribution with rate μ. The number of service channels, represented by c, corresponds to the number of robots or human workers handling order processing. The probability of an order experiencing a delay in fulfillment is given by the Erlang-C formula:

$$P_w = \frac{\frac{(c\rho)^c}{c!}\frac{1}{1-\rho}}{\sum_{k=0}^{c-1}\frac{(c\rho)^k}{k!} + \frac{(c\rho)^c}{c!}\frac{1}{1-\rho}} \tag{7.4}$$

where $\rho = \frac{\lambda}{c\mu}$ represents the system utilization. Amazon optimizes the number of service channels to minimize queueing delays while balancing labor costs [15].

Last-mile delivery, one of the most resource-intensive components of e-commerce logistics, is optimized using Vehicle Routing Problem (VRP) formulations. The objective is to minimize the overall delivery cost by determining the optimal routes for delivery vehicles, taking into account demand constraints and vehicle capacities. The VRP formulation can be expressed as:

$$\min \sum_{i\in V}\sum_{j\in V} c_{ij} \cdot x_{ij} \tag{7.5}$$

subject to:

$$\sum_{j\in V} x_{ij} = 1, \forall i \in V \tag{7.6}$$

which ensures that each customer is served exactly once, and

$$\sum_{i\in V} d_i \cdot x_{ij} \leq C, \quad \forall j \in V \tag{7.7}$$

which enforces the vehicle capacity constraint. Here, c_{ij} represents the transportation cost between nodes i and j, d_i is the demand at location i, and C is the vehicle's maximum load capacity. Amazon applies machine-learning-enhanced VRP solvers to dynamically adjust routes based on real-time traffic conditions, ensuring cost-efficient and timely deliveries [15].

Amazon's automated e-commerce distribution network leverages a combination of robotics, AI, queuing models, and network optimization techniques to achieve superior efficiency. By modeling its logistics operations through graph-based representations, queuing theory, and VRP formulations, Amazon continuously enhances its ability to fulfill orders quickly and at reduced operational costs. The integration of real-time data analytics, predictive algorithms, and automation systems has positioned Amazon as a benchmark in modern logistics. The ability to anticipate demand fluctuations, adapt to supply chain disruptions, and optimize last-mile delivery ensures that Amazon remains at the forefront of e-commerce logistics innovation.

The e-commerce company's distribution network can be configured as a small-world network. Apply spectral gap maximization to optimize warehouse locations and transportation links, reducing average delivery times and enhancing customer satisfaction is a smart application enhancing the engineering perspective [29].

China's e-commerce landscape is dominated by giants like JD.com and Alibaba, both of which have revolutionized their distribution networks through advanced automation and intelligent logistics systems.

JD.com has been at the forefront of integrating automation into its logistics operations. A notable example is its "Asia No.1" logistics park, which stands as one of the world's largest intelligent logistics facilities. This state-of-the-art warehouse boasts integrated warehousing and sorting capabilities, significantly enhancing operational efficiency and reducing delivery times. The facility employs a range of automation tools, including miles of conveyors, cross-belt sorting systems, six-axis robotic arms, Autonomous Mobile Robots (AMRs), and Container Transport Unit (CTU) robots, all working in unison to streamline the sorting and distribution processes (see Figure 7.2).

Such innovations have enabled JD.com to handle massive order volumes with minimal human intervention, exemplifying the future of automated logistics.

Similarly, Alibaba has made significant strides in optimizing its logistics network through its logistics arm, Cainiao Network. Established in 2013, Cainiao has evolved into a global logistics powerhouse, leveraging big data, artificial intelligence, and the Internet of Things (IoT) to enhance efficiency across the supply chain (see Figure 7.3). One of its key initiatives includes the development of a smart logistics solution that simplifies warehousing workloads with automated operations and a comprehensive management system.

This solution accelerates location-based services with high-accuracy address processing capabilities, ensuring that logistics operations remain on schedule and responsive to customer demands. By integrating these advanced technologies, Alibaba aims to deliver packages within 24 hours across China and within 72 hours globally, setting new standards in e-commerce logistics.

Figure 7.2 Example for CTU robots at JD. Source: courtesy of https://jd.com

7.2.1 GREEN ROUTING IN TRANSPORTATION

Companies like UPS implemented *green routing algorithms* to minimize emissions, leveraging multi-objective optimization for sustainability. UPS has demonstrated a strong commitment to sustainability by integrating advanced technologies to optimize its transportation routes to the fullest, thereby reducing environmental impact and increasing operational efficiency. One such effort in this regard is the On-Road Integrated Optimization and Navigation (ORION) system, which employs big data and artificial intelligence to streamline delivery operations.

ORION, founded in 2013, was established to optimize UPS drivers' delivery routes by analyzing a staggering amount of data points, including package status, customer preference, traffic flow, and fueling behavior. The platform processes these data points to produce the most optimal routes available, with the aim of minimizing left-hand turns, lowering idle time, and maintaining overall low fuel consumption. By 2017, ORION had been implemented on 55,000 routes and contributed significantly to UPS's sustainability initiative [6].

The implementation of ORION has led to substantial reductions in greenhouse gas emissions. By optimizing routes and decreasing fuel consumption, UPS has

Figure 7.3 Areas of development for logistic solutions at Cainiao. Source: courtesy of https://www.cainiao.com

been able to cut emissions by millions of metric tons annually. The system has improved delivery efficiency, resulting in significant cost savings. By the end of 2013, ORION had optimized 10,000 delivery routes, with plans to extend this optimization to 55,000 routes by 2017, leading to substantial reductions in transportation costs. As of 2023, UPS continues to leverage its ORION system to enhance operational efficiency and sustainability. ORION utilizes advanced algorithms to process complex datasets, including real-time traffic patterns, weather conditions, vehicle specifications, and delivery commitments, enabling the optimization of delivery routes.

UPS operates in over 200 countries and territories, serving approximately 1.6 million shipping customers and 10.2 million delivery recipients daily. In 2023, UPS handled an average daily volume of 22.3 million packages and documents, underscoring the extensive application of ORION in managing large-scale logistics operations. The company's operating profit stood at USD 9.1 billion, with an adjusted operating profit of USD 9.9 billion, marking a 28.7% decrease. In late 2024, UPS announced plans to reduce its business volume with Amazon, its largest customer, by 50% over the next 18 months. This strategic move aims to redirect resources toward more profitable sectors, such as healthcare logistics, with a target to increase revenue in this segment to USD 20 billion by 2026.

Through the continued implementation of ORION and strategic business adjustments, UPS demonstrates its commitment to optimizing logistics operations, enhancing efficiency, and maintaining a competitive edge in the evolving transportation industry.

The Energy-efficient vehicle routing problem can be formulated as a graph-grounded problem, as described by eq. 7.5 to eq. 7.7. In addition, tardiness and energy constraints need to be defined.

Time Windows Constraint: Each delivery must be completed within a specified time window $[T_j^{start}, T_j^{end}]$:

$$T_i + t_{ij} \cdot x_{ij} \leq T_j \forall (i, j) \in E. \tag{7.8}$$

where t_{ij} is the estimated travel time.

Energy Efficiency Constraints: The fuel consumption per route segment can be modeled using a speed-dependent fuel efficiency function:

$$c_{ij} = \alpha \cdot d_{ij} + \beta \cdot v_{ij}^2 + \gamma \tag{7.9}$$

where d_{ij} is the distance, v_{ij} is the vehicle speed, and α, β, γ are fuel efficiency parameters.

To optimize fuel consumption dynamically, AI-driven Reinforcement Learning (RL) techniques can be applied.

A Deep Q-Network (DQN) can be trained to adjust routes based on Traffic conditions, Real-time weather data, Vehicle load distribution and Delivery time constraints.

For RL the state space includes road congestion levels, fuel efficiency metrics, and time constraints. The action space consists of possible route choices. The reward function is designed to minimize the fuel cost while meeting delivery deadlines.

$$R(s,a) = -\sum_{(i,j)} c_{ij} \cdot x_{ij} - \lambda \sum_{j} (T_j - T_j^{end}, 0) \tag{7.10}$$

where λ is a penalty factor for exceeding the time window.

In the realm of Reinforcement Learning (RL), an agent learns to make optimal decisions through interactions with an environment. The objective is to maximize cumulative rewards over time by choosing the best possible actions in given states. A Deep Q-Network (DQN) is a neural network-based RL algorithm that helps to dynamically adjust delivery routes by predicting the best decision in each traffic scenario while minimizing fuel consumption and maintaining on-time deliveries. When applied to UPS route optimization case, the agent represents the route selection system, and the environment consists of the real-world delivery network, including traffic conditions, road types, fuel consumption rates, and time constraints.

DQN is built on the foundation of Q-learning, which is a technique used to find the best action to take given a specific state. In classical Q-learning, an agent maintains a Q-table that stores expected rewards for taking an action in a specific state. The Q-value is updated iteratively using the Bellman equation:

$$Q(s,a) = Q(s,a) + \alpha \cdot (r + \gamma \cdot max_{a'} Q(s',a') - Q(s,a)) \tag{7.11}$$

where:

- $Q(s,a)$ represents the expected reward for taking action a in state s.
- α is the learning rate, controlling how much new information influences the existing Q-value.
- r is the immediate reward received after taking action a.
- γ is the discount factor, which determines how much future rewards are considered.
- $\max_{a'} Q(s',a')$ is the maximum future Q-value, assuming the best next action is chosen.

However, traditional Q-learning struggles in complex environments like UPS's logistics system, where there are thousands of possible routes, dynamic traffic updates, and real-time delivery constraints. This is where Deep Q-Networks (DQN) come in.

Instead of storing all possible $Q(s,a)$ values in a Q-table, which becomes impractical for large-scale problems, a Deep Neural Network (DNN) is used to approximate the Q-function. This enables the system to generalize and predict the best actions without storing every possible scenario.

The DQN model takes as input the state representation, that is, the current delivery environment, including the current vehicle location and package destinations, the traffic congestion level on different roads, and the estimated fuel consumption per route, as well as the estimated delivery time remaining. It also considers the action choices as the available route options the vehicle can take next. Also, the predicted Q-values: The network outputs an estimate of future rewards for each possible action. The action with the highest Q-value is selected. This way, DQN allows UPS to choose the most fuel-efficient and timely routes dynamically, without needing explicit human intervention.

Training a DQN involves several key steps:

- Exploration and Exploitation: The model initially explores different routes randomly to collect diverse data, then gradually shifts toward exploiting the best-known paths.
- Replay Memory: Instead of learning from recent actions alone, the DQN stores past experiences in a memory buffer and samples from them randomly. This prevents the network from overfitting to recent events.
- Target Network: To stabilize learning, DQN maintains a second, slowly updating "target" network that helps prevent sudden changes in predictions.
- Loss Function Optimization: The DQN minimizes the difference between predicted Q-values and actual Q-values using backpropagation and gradient descent.

Through these techniques, the system learns progressively better routing strategies, balancing fuel savings, delivery times, and real-time traffic changes.

7.2.2 RESILIENCE IN PHARMACEUTICAL SUPPLY CHAINS

Aspen Pharmacare, a prominent South African pharmaceutical company, offers an extensive portfolio of cost-effective, high-quality, niche manufacturing capabilities across a global network. They have a range of scalable production capabilities and capacities aligned with their current and future commercial objectives, including injectables, oral solid and semisolid, liquids, and active pharmaceutical ingredients (APIs). They operate 23 manufacturing facilities across 15 sites, holding international manufacturing approvals from some of the most stringent global regulatory agencies, including the United States Food and Drug Administration, the Australian Therapeutic Goods Administration, and the European Directorate for the Quality of Medicines. The company operates with an established business presence in approximately 56 countries/regions and supply products to more than 150 countries, and its business performance can be seen in Figure 7.4.

In 2020, Aspen Pharmacare faced a massive supply chain dilemma. Worldwide demand for dexamethasone, a corticosteroid that was found effective in treating critical COVID-19 patients, abruptly spiked. Aspen, being one of the primary producers, had to act swiftly on this unexpected spike to ensure timely delivery of the medication. Recognizing the urgency, Aspen leveraged its robust supply chain infrastructure to scale up production. Aspen's prior investment in high-containment facilities at Gqeberha (previously Port Elizabeth) was instrumental in this effort. Through these scalability-designed fast-tracked facilities, Aspen managed to meet the heightened

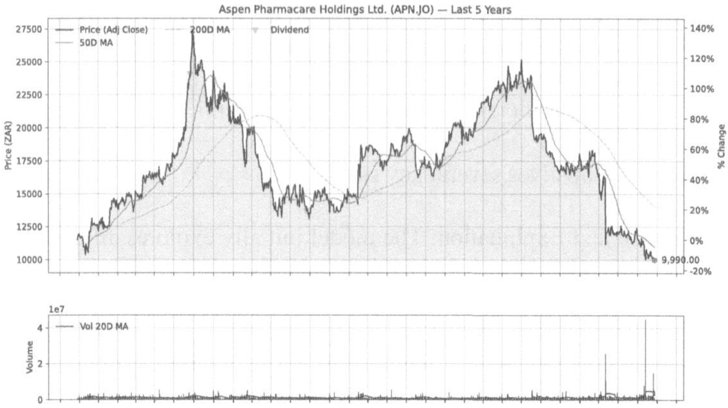

Figure 7.4 Performance for Aspen Pharmacare

demand without any compromise on quality and safety standards. In addition, Aspen's supply chain management was proactive, with close interaction with raw material suppliers to guarantee a continuous flow of required components. The company's strategic move to have diversified sourcing channels hedged against possible disruptions, demonstrating a high level of resilience.

Aspen Pharmacare's achievement in coping with the demand surge for dexamethasone in the COVID-19 pandemic was largely attributed to its well-developed resilience system. The company leveraged various determinants to measure and maintain supply chain resilience and hence avoid interruptions from impacting production or delivery to any significant degree, such as:

- *Supply chain agility* was one of the key elements of resilience. Aspen had to reduce lead times considerably so that additional production could be brought on-stream in a rush. Its production processes were flexible, and this was crucial, as it could shift resources efficiently toward dexamethasone production without requiring major reconfigurations. This flexibility was not restricted to internal operations, as the company worked with suppliers to accelerate the supply of raw materials. The ability of these suppliers to respond to Aspen's needs became a key measure of resilience.
- *Redundancy and diversification* in the supply chain also provided additional backup. Aspen followed a multi-supplier policy, which meant that it was not too reliant on a single source of major raw materials. The diversification minimized risks, preventing the kind of bottlenecks that would have hindered production. In addition, the firm had adequate inventory buffers, so even if there were short-term supply shortages, production would not be significantly disrupted. One of the main reasons Aspen was able to weather the storm was that it had invested in high-containment facilities in Gqeberha, which enabled an instant ramp-up in production when demand surged.

- *Risk monitoring* and *disruption recovery* were also core to the company's success. Aspen had mechanisms for identifying potential risks in its supply chain, which allowed it to be proactive rather than reactive. Monitoring the duration taken to recover from disruptions was essential, as it allowed the company to track its ability to return to normal production speedily after unexpected events. Minimizing production downtime was another essential parameter, allowing capacity to be always optimized.
- *Supply chain visibility* played a significant role in helping the business remain resilient. The company had end-to-end visibility of its entire supply base, such that it could identify potential issues before they turned into problems. Transparent information on raw materials availability, production schedules, and distribution networks ensured that Aspen could respond in a timely and informed manner. Through this data-driven approach, it was possible to successfully employ forecasting models to anticipate downturns in demand and to adjust operations accordingly.
- Equally important was Aspen's *collaboration with the supply chain partners*. Effective working with suppliers, regulators, and distributors ensured free movement of materials and finished goods. The company had to keep up with regulatory requirements in good time, particularly noting the urgency in distributing dexamethasone to COVID-19 patients. Effective communication with government authorities and medical professionals ensured effective distribution of the drug to the places where it was most needed.

The time taken to resume normal supply chain operations (RTO) after an unexpected event (such as increased demand) was adopted as resilience metric. Having safety stock for critical raw materials and finished products played a role in sustaining production without immediate disruptions. Aspen needed to track raw material availability, production schedules, and distribution logistics to ensure steady product flow. Therefore, the use of forecasting models to anticipate demand surges and adjust supply chain operations accordingly was a keystone engineering process.

Through such resilience steps, Aspen was able to raise output without facing imminent shortages of material. It ensured that dexamethasone remained in supply even during the peak of the crisis, solidifying its reputation as a reliable pharmaceutical supplier even more. The crisis not only tested Aspen's resilience but also improved its ability to recover from future interruptions. The company released a better supply chain that was better able to manage future demand surges and could more readily safeguard against threats. Through these concerted efforts, Aspen not only met the worldwide demand surge but also reinforced its reputation as a reliable pharmaceutical supplier in a time of necessity.

This case demonstrates the merit of investing in infrastructure and strategic planning to enhance supply chain resilience in the pharmaceutical industry, but also the strong interconnections between engineering and logistics.

Using *percolation theory*, pharmaceutical companies assessed vulnerabilities in drug distribution networks to prevent shortages during crises. Percolation theory, originally developed in statistical physics to describe connectivity in disordered systems, has been widely used in network theory, epidemiology, and supply chain resilience analysis (see Sections 6.1.5.2 and 6.1.5.3 for more details).

As depicted in Section 6.1.5.1, to quantify resilience, people define the average shortest path length ℓ, which describes how efficiently resources (such as raw materials) can travel through the network:

$$\ell \sim \frac{\log N}{\log\langle k\rangle} \tag{7.12}$$

where N is the number of nodes. A well-connected network, such as Aspen's diversified supplier network, maintains a small ℓ, ensuring that raw materials and finished drugs are distributed efficiently even under partial disruptions.

In Aspen's context, it provides a useful framework to analyze how disruptions propagate through the supply network and how critical thresholds determine whether the system remains functional or collapses. Aspen's pharmaceutical supply chain can be modeled as a complex network, where raw material suppliers, manufacturing plants, distributors, and hospitals act as nodes, while transportation and logistical pathways serve as the edges connecting them. Percolation theory helps assess how supply chain resilience depends on the connectivity of these nodes. If too many critical nodes (such as key suppliers of raw materials or distribution hubs) fail, the system could reach a percolation threshold, leading to a cascading failure where disruptions spread uncontrollably. What percolation theory ultimately reveals is that resilience is not just about having strong individual components but about maintaining connectivity across the entire system.

7.2.3 IOT-ENABLED LOGISTICS TRACKING

IoT-enabled logistics has significantly enhanced tracking, and the business value has been tremendous. A case in point is the collaboration between Polymer Logistics and Hoopo, an Israeli geolocation company. Together, they introduced IoT tracking within the supply chain, enabling the possibility to run logistics operations effectively by creating data visibility on the location of assets. It stops bottlenecks, allows asset maintenance, and offers real-time tracking through cloud software and interoperable hardware.

Before this collaboration, Polymer Logistics, a global leader in reusable packaging and pooling services now owned by Tosca Ltd. from 10/16/2019, had established itself as an innovative provider of "One-Touch / Retail Returnable" packaging solutions, serving retailers worldwide [3]. Their focus was on enhancing supply chain efficiency through reusable transport packaging and retail merchandising systems. However, like many in the industry, they faced challenges in tracking assets across complex supply chains, leading to inefficiencies and increased operational costs. In February 2018, Polymer Logistics embarked on a transformative partnership with Hoopo, a technology company specializing in low-power, wide-area Internet of Things (IoT) solutions.

Hoopo, on the other hand, had been redefining geolocation technology with its high-accuracy, low-power capabilities via IoT. Their expertise lay in creating

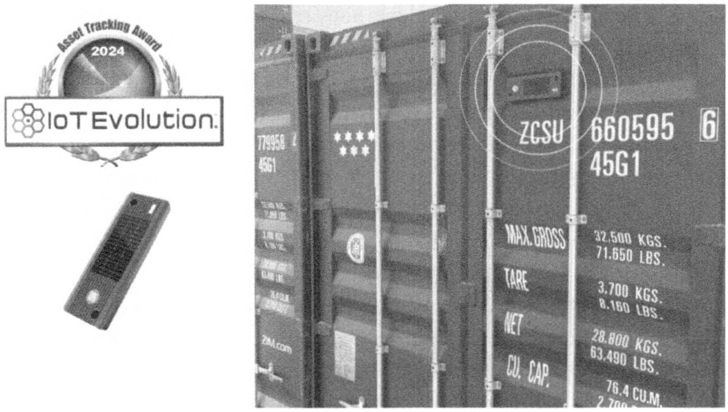

Figure 7.5 IoT hoopoSense Solar device and a typical location in a container. Source courtesy of https://hoopo.tech/

bespoke end-to-end IoT tracking devices that digitized delivery assets, enabling companies to monitor the status and location of these assets in real-time without relying on traditional GPS systems.

Hoopo's flagship product, hoopoSense Solar (see Figure 7.5), is a solar-powered tracking device boasting a remarkable battery life of over 12 years, ensuring long-term, maintenance-free operation. This device provides real-time visibility and traceability, enhancing operational efficiency for businesses. The hoopoSense Solar tracker is designed to withstand harsh environments, featuring a rugged and durable build with an IP67 rating. It offers wide global coverage through support for LTE-M, NB-IoT, and 2G networks, ensuring reliable connectivity in diverse locations. Installation is swift, taking approximately 90 seconds, minimizing downtime during deployment. Hoopo's technology leverages multiple location and connectivity protocols, adapting to various operational needs to maintain accurate and reliable tracking. The system includes features such as tamper detection, door open detection, impact detection, and load monitoring, providing comprehensive insights into asset status. Additionally, the platform utilizes smart transmission methods to optimize power efficiency, ensuring prolonged device operation without frequent maintenance.

Through the incorporation of IoT devices and sensors, businesses are able to garner real-time information about different key elements of logistics processes, such as the location of vehicles, traffic, and the status of transported items. With this information, exact route planning, resource allocation, and proactive decision-making become possible, which translates to optimized operations and better cost-efficiency.

IoT implementation in logistics is not an upgrade but a game-changer. With the IoT market in logistics globally expected to reach USD 100 billion by 2025, the technology is fast becoming a transformative force. By leveraging IoT technology, organizations can achieve visibility, routes can be optimized, and timely delivery of goods can be assured.

Route optimization through IoT leverages real-time data collection, advanced analytics, and predictive algorithms to improve efficiency in logistics and supply chain operations. Traditionally, route planning relied on historical data and static scheduling, which often failed to account for dynamic variables such as traffic congestion, weather conditions, vehicle conditions, and delivery constraints. With IoT, vehicles, shipments, and infrastructure elements are interconnected through sensors and GPS-enabled tracking devices, continuously transmitting data to centralized platforms that process and analyze this information.

At the root of this optimization is IoT's ability to provide real-time visibility into vehicle location and status. Sensors installed in shipping containers or delivery vehicles monitor metrics such as fuel consumption, engine health, and temperature control for perishables. Should a vehicle encounter unexpected delays due to traffic, road closures, or breakdowns, the system immediately detects the problem and suggests rerouting. Advanced algorithms not only look at the most direct route possible but also consider fuel consumption, driver fatigue, and delivery priority to ensure rerouting decisions align with business goals and regulatory compliance. One key feature of smart IoT-based route optimization is integrating real-time traffic data from a range of sources, including GPS, road sensors, and external databases. This enables logistics operators to foresee possible bottlenecks and dynamically rearrange routes even before delays are encountered. When a delivery is routed through a severely congested route, for example, the system can recommend a quicker option by computing traffic density and arrival times. Furthermore, IoT devices placed in warehouses and distribution centers provide real-time status on shipment readiness, allowing vehicles to be dispatched the moment merchandise is ready to ship, reducing downtime and improving overall efficiency.

Beyond mere tracking, IoT-powered route optimization incorporates machine learning algorithms that dynamically recalibrate route suggestions based on previous performance and environmental factors. By analyzing historical delivery times, fuel consumption patterns, and chronic delays in certain regions, these systems become progressively smarter, learning the routes that ensure best-in-class performance under varied conditions. By predicting logistics in advance, this enables companies to proactively manage logistics, hedging against the dangers of late deliveries and exorbitant costs. The second critical component is vehicle-to-vehicle (V2V) and vehicle-to-infrastructure (V2I) communication, whereby IoT devices in trucks communicate with other vehicles and roadside sensors. This networked system enhances safety and efficiency by enabling vehicles to coordinate speeds, maintain optimal following distances, and even prepare for risks such as sudden braking by another vehicle. These adjustments not only enhance fuel efficiency but also overall fleet longevity by reducing wear and tear on vehicles.

IoT sensors monitor shipments in real-time, providing temperature and humidity tracking in cold-chain logistics. In industries that require temperature-sensitive deliveries, such as pharmaceuticals and food distribution, IoT-enabled route optimization becomes even more valuable. Sensors inside refrigerated trucks monitor temperature and humidity, ensuring that perishable goods remain within required conditions. If an issue is detected, the system can reroute the truck to the nearest service point or adjust the delivery sequence to prevent spoilage. This level of intelligence ensures that product quality is maintained while minimizing waste, contributing directly to business profitability.

The integration of IoT in route optimization does not stop at delivery vehicles. It extends to last-mile logistics, where smart tracking devices on parcels communicate with delivery personnel, allowing customers to receive real-time updates on estimated arrival times. This increases customer satisfaction while reducing missed deliveries, which are costly to businesses. Furthermore, advanced route optimization process ensures that last-mile couriers take the most efficient paths, avoiding unnecessary stops and minimizing fuel consumption. Through continuous data exchange, intelligent learning, and predictive analytics, IoT transforms route optimization into a dynamic, self-improving system that enhances operational efficiency, lowers costs, and improves customer service. By making logistics networks more responsive and adaptive, businesses gain a competitive advantage while reducing environmental impact through smarter fuel consumption and reduced emissions. The ability to process vast amounts of real-time data in a meaningful way is what makes IoT-driven route optimization a powerful and indispensable tool for modern logistics.

In summary, IoT-based logistics solutions provide real-time tracking and monitoring that leads to route optimization, efficient use of resources, and enhanced customer satisfaction. These advantages translate to easy operations, increased cost-effectiveness, and a competitive edge in the competitive logistics market.

7.2.4 SUSTAINABLE REVERSE LOGISTICS

GreenTech Solutions, based in the UK, but working globally, is worldwide producer of electronics. The company had been committed to sustainability for a long time, yet its reverse logistics were becoming increasingly complex by the day.

With millions of electronic items returned every year—ranging from smartphones and laptops to batteries and smart home devices—the company was faced with the following challenge: how to efficiently manage returns while maximizing component reuse, minimizing environmental impact, and complying with stringent e-waste regulations in dozens of countries. At the heart of the problem was the unpredictability of returns. Products were being returned by consumers via various channels—warranty claims, trade-in programs, and independent recycling programs—creating a network of collection points, sorting centers, repair shops, and recycling centers. The cost of transporting these products, in many instances great distances, needed to be balanced against sustainability goals, low carbon emissions while also controlling operating costs. Decision-making was further complicated due to the different possible destinies of each returned product: some could be refurbished and resold, some could be dismantled for valuable components, and some needed to be disposed of in an environmentally friendly fashion.

That is where GreenTech Solutions turns to advanced mathematical modeling, employing graph theory and Petri nets in order to optimize its reverse logistics processes [13]. Their engineering and data science teams created a dynamic, weighted graph representation of their entire reverse supply chain. The nodes of the graph were the critical points in the logistics flow—customer drop-off locations, collection centers, repair facilities, recycling facilities, and secondary markets—while edges between the nodes accounted for transportation lanes, costs, and carbon footprint [1]. Using shortest-path algorithms such as Dijkstra's and A*, the company would be able to compute the most economical and sustainable path for each device that was returned, with transport decisions in real-time being optimized by factors such as fuel costs, available storage space, and emissions reduction pledges.

Petri nets are a mathematical and graphical modeling technique to analyze, model, and simulate discrete event systems, particularly concurrent, asynchronous, parallel, and distributed processes. Petri nets were originally conceived in the 1960s by Carl Adam Petri. Petri nets enable systematic description of dynamic systems with multiple states and transitions occurring simultaneously and are especially applied in logistics, manufacturing, computer networks, and workflow management. Essentially, Petri nets consist of three simple components: places, transitions, and directed arcs. Places are employed to represent states or conditions, transitions represent events or actions that alter the states, and directed arcs connect places to transitions to indicate how tokens move through the system. Token movement through places and transitions defines system behavior, modeling event sequences, decisions, and interdependencies between different operations. Yet another of the Petri net's greatest strengths is its capability for representing concurrency. Unlike more traditional flowcharts or decision trees, which impose a rigid sequential ordering, Petri nets allow for several transitions to be fired at once or in parallel, echoing the true parallelism of much real-world activity. This makes them particularly well-suited for modeling logistics and supply chains, where various tasks—such as transportation, sorting, repair, and recycling—can happen in parallel but still be dependent upon one another [22].

Mathematically, the formal definition of a Petri net is a bipartite graph, i.e., places and transitions constitute two distinct sets, and directed arcs connect only elements between two different sets. The bipartite property avoids mixing of conditions and events. That is, Petri Net is a net understood as a tuple $N = (P, T, F)$ where P are the sets or places and T are the transitions or directed arcs. In addition $F \subseteq (P \times T) \cup (T \times P)$ and a marking set $M : P \to Z$, with Z a countable set. and W as a multiplicity setting $W : F \to Z$. Every transition is triggered whenever all its input places contain the number of tokens it requires, and when it fires, it consumes tokens from the input places and generates new tokens in the output places. This simulates causality in the physical world, where certain conditions must be met before an event can occur, and the event subsequently causes other conditions. In addition to basic Petri nets, there are other extensions such as timed Petri nets, stochastic Petri nets, and colored Petri nets that provide additional features to study complicated phenomena. Timed Petri nets include delays in transitions to make performance analysis in time-critical domains like logistics and production planning possible. Stochastic Petri nets include probabilities, enabling expression of uncertainty and variability, significant in complex decision-making processes such as reverse logistics, where product return rates and processing times are different. Colored Petri nets enable tokens to bear distinctive attributes or identifiers, useful in scenarios where objects must be distinguished—such as product categories or types of customers.

The analytic strength of Petri nets surpasses visual representation because they are also analyzable mathematically by reachability graphs, incidence matrices, and invariant analysis. Reachability graphs examine all states that the system can be in, guaranteeing completeness in analyzing dynamic behavior. Incidence matrices give algebraic information about how transitions influence the token distribution among places, allowing structural analysis of the system. Invariant analysis, i.e., place invariants and transition invariants, is utilized to verify system properties such as resource conservation or correctness of a workflow.

Petri nets are particularly valuable in supply chain optimization and logistics because they offer a means to detect inefficiencies, identify bottlenecks, and ensure correctness of workflow execution. They provide a mathematically sound approach to process synchronizations, deadlocks, and throughput bottlenecks analysis, which is critical for industries with complex networks of material flow, service operations, and decision-making nodes. Lastly, Petri nets are both a conceptual and computational tool that creates a loop between system design, simulation, and formal verification. Their ability to handle concurrency, randomness, and resource constraints makes them a critical tool in sectors where many interdependent processes develop dynamically, like in sustainable reverse logistics operations, where returns that arrive need to be valued, repaired, refurbished, recycled, or discarded in a way that keeps costs low and impacts on the environment small.

Streamlining the logistics network was only one side of the coin, though. The business needed a way of understanding and streamlining the inner workings of its processing plants, through which returned products passed for inspection, sorting, repair, and disassembly. To tackle this, the operations researchers at GreenTech turn to Petri nets. Each device, upon entering the reverse logistics process, was a token in a complex workflow model. Places in the Petri net represented different processing steps—initial diagnostics, component-level testing, refurbishing, recycling, and final disposal—while transitions modeled decision points and operation constraints. With this model, the company could simulate numerous return scenarios, uncovering bottlenecks in its processes, anticipating capacity constraints at certain of its facilities, and scheduling resources dynamically to prevent delays. One of the most important insights was realized in modeling the repair decision process. By modeling failure rates and processing times in a stochastic Petri net, GreenTech discovered that certain categories of returns—especially older models of smartphones—were occupying repair facilities even though they had very low resale value. The simulations showed that redirecting these devices directly to specialized recycling facilities would not only free up capacity but also increase material recovery rates by 15% and reduce reliance on newly mined raw materials.

The integration of these technologies provided tangible gains. The transportation cost was reduced by nearly 20% due to route planning optimization, and carbon emissions across the reverse logistics chain were reduced by 25%. Decision-making for repair and recycling facilities was made far more efficient, with high-value returns being prioritized and low-value, high-impact waste being disposed of sustainably. Further, the use of graph-based predictive analytics helped GreenTech predict seasonal return patterns and dynamically scale their logistics operation to avoid bottlenecks during peak periods.

Eventually, GreenTech Solutions' commitment to sustainability was no longer just a corporate goal—it was now embedded in every data-driven decision, every optimized route, and every streamlined process. By harnessing the power of graph theories (see Chapter 6 and Petri nets), they had transformed their reverse logistics operations into a model of efficiency, circular economy leadership, and environmental responsibility, setting a new standard for the industry.

7.3 CONCLUSION

The applications examined within this chapter exemplify how graph and mathematical theories form the basis of improving intricate decision-making processes. Essentially, these methods offer the structural and analytical tools needed to model, analyze, and improve logistical systems wherein uncertainty, concurrency, and resource allocation are important determinants. Graph theory, which is capable of representing networks of interconnected elements, is most useful in reverse logistics, where different locations such as collection points, repair shops, recycling facilities, and resale markets interact with one another dynamically. By modeling transportation routes as weighted graphs and employing shortest path algorithms, businesses are able to cut costs, reduce environmental impacts, and optimize product flows so that returned products reach the most appropriate destination in an efficient manner.

Petri nets, on the other hand, offer a formal mathematical representation of the product flow in reverse logistics systems, enabling engineers to map the various decision points, processing phases, and potential bottlenecks. Their bipartite structure enables precise modeling of states and transitions, which enables organizations to model alternative logistical scenarios and forecast the consequences of variations in return volumes, processing times, and resource availability. The ability to integrate timing, probability, and attribute into Petri net models broadens their applicability and makes them a priceless asset in determining the stochastic behavior of reverse logistics processes when product returns are unpredictable.

Engineering brings these mathematical models into productive practice. Crossing math tools with practical constraints, engineers develop algorithms, simulations, and optimization platforms to guide decision-making. Using machine learning, operational research, and systems engineering, they improve these models to incorporate unforeseen factors such as fickle consumer behavior, regulatory changes, and technology evolution. Engineering provides that decision-making is not merely mathematically correct but adjusted to the characteristics of business processes, leading to sustainable, cost-effective, and efficient reverse logistics solutions. The interplay of decision sciences, engineering, and mathematics thus gives rise to a strong ecosystem wherein intricate logistical challenges can be approached systematically, advancing innovation and sustainability in modern-day supply chains.

In addressing efficiency and robustness in supply chain management, more levels of mathematical modeling and engineering principles need to be incorporated into the decision-making process. Efficiency within a supply chain involves optimizing resources, lowering costs, and limiting delays, while robustness involves ensuring functionality and performance in the face of uncertain or disruptive conditions. Both require a more intensive use of graph theories, optimization methods, and resilience engineering.

Efficiency is typically improved by network optimization, whereby advanced graph theory concepts such as minimum spanning trees, flow networks, and centrality measures are applied to identify cost-reducing and time-reducing paths for material flow. Multi-objective optimization methods, such as linear programming, integer programming, and metaheuristic methods (e.g., genetic algorithms or particle swarm optimization), allow firms to balance cost savings with sustainability goals. Additionally, dynamic graph algorithms help in real-time routing decisions, altering transportation routes based on changing constraints like fuel costs, demand variability, and storage capacities.

Robustness, on the other hand, requires supply chains to withstand disruptions in the form of supply disruption, transportation breakdown, or sudden variations in return volumes. Resilience modeling using stochastic Petri nets introduces probabilistic elements into system workflows, allowing firms to simulate worst-case scenarios and put mitigation strategies in place. Graph-based redundancy analysis, using concepts such as k-connectivity and network resilience metrics, ensures that there are alternate paths in the logistics network, such that the system does not collapse when certain nodes or paths are impacted. Additionally, game theory and reinforcement learning frameworks allow supply chains to react dynamically to market fluctuations and exogenous disruptions.

From an engineering perspective, real-time monitoring systems and digital twins enhance both robustness and efficiency. By integrating Internet of Things (IoT) sensors, machine learning models, and predictive analytics, organizations are able to make data-driven adjustments to their supply chain in response to anomalies. Agent-based modeling (ABM) and Monte Carlo simulations provide additional robustness by modeling how different agents—suppliers, transportation providers, customers, and so on—behave under uncertainty. The application of this advanced mathematical and engineering approach ensures not only that supply chains are operated with maximum efficiency under ordinary conditions but also that they are resilient against unexpected disruptions. This dual focus on decision-making strengthens companies' abilities to provide sustainability and cost-effectiveness while safeguarding against potential breakdowns in their reverse logistics and supply chain networks as a whole.

8 Emerging Trends and Advanced Topics

8.1 INTRODUCTION

The goal of this chapter is to observe trends and connected topics and technologies that can be integrated and used to further improve logistics operations around the world. Based on current technological investments, as well as market needs, several innovative fields and future trends will have a significant impact on logistics, driven by increasing technological change, increasing demands for sustainability, and the need for resilience in global supply chains. The most significant fields will be those that enable automation, optimization of decision-making, and optimization of efficiency while ensuring robustness and sustainability. The most significant areas that will shape the future of logistics include Artificial Intelligence and Machine Learning, Digital Twins and Simulation, Blockchain for Supply Chain Transparency, Quantum Computing for Optimization, and Autonomous Logistics Systems.

AI and machine learning will have an even more widespread impact on logistics, particularly in predictive analytics, demand forecasting, and real-time optimization. Advanced AI models will streamline logistics operations through learning from huge amounts of data, allowing firms to anticipate supply chain disruptions, predict shifts in demand, and dynamically reset inventory levels. Reinforcement learning models based on AI will also improve route planning and warehouse automation, reducing costs, and improving service levels. The integration of AI and graph neural networks (GNNs) will lend additional strength to the analysis of complex supply chain networks, making logistics operations more resilient and adaptive.

Digital twins will revolutionize supply chain management through real-time virtual representations of logistics networks. These virtual copies, powered by IoT sensors and big data analytics, will enable companies to simulate different scenarios, assess risk factors, and test optimization strategies in the virtual environment before they are executed in the physical world. With the help of digital twins, supply chain managers will be able to create self-adaptive supply chains that respond in real-time to changing circumstances, such as disruptions caused by extreme weather, political instability, or economic uncertainty. The integration of digital twins with Petri nets will enhance their applicability to represent complex concurrent logistics processes, allowing firms to identify bottlenecks and optimize workflows in real time.

Blockchain technology will play a more critical role in improving transparency, security, and traceability in logistics chains. The decentralized nature of the blockchain will deter fraud, track shipments more effectively, and impose adherence to sustainability regulations. With smart contracts, companies will automate transactions such as payments, customs clearance, and compliance verification to

DOI: 10.1201/9781003627906-8

remove bureaucratic redundancies and speed up international transactions. Graph-based ledger analytics will be instrumental in detecting fraud and trustworthiness in multi-level supply chains.

Quantum computing will revolutionize supply chain optimization problems that are currently computationally infeasible. Classical optimization methods are often stymied by the combinatorial scale of large logistics networks, but quantum algorithms, such as quantum annealing, will provide exponential speedups for routing, scheduling, and warehouse optimization problems. In particular, quantum-inspired heuristics will improve last-mile delivery routing, lowering costs and emissions while ensuring on-time delivery.

Autonomous logistics, including driverless trucks, robotic warehouses, and drone deliveries, will revolutionize efficiency and cost-effectiveness. With greater adoption of 5G and edge computing, real-time orchestration of autonomous vehicles and robotics will become feasible, reducing the need for human intervention in logistics processes. AI-powered swarm intelligence algorithms will be applied to orchestrate swarms of autonomous delivery drones and trucks, creating adaptive and decentralized logistics networks that can function even under unexpected disruptions.

The principles of sustainability and circular economy will also define the future of logistics, with an increasing emphasis on carbon-neutral supply chains, energy-efficient transportation, and waste reduction through advanced recycling technologies. AI-driven multi-objective optimization models will help organizations optimize cost, efficiency, and environmental impact; address compliance with stringent regulations; and shift consumer sentiment in favor of sustainability. However, these objectives need to be articulated against higher levels of transparency between stakeholders.

The convergence of these emerging disciplines—AI, digital twins, blockchain, quantum computing, autonomous logistics, and sustainability-focused optimization—will transform logistics into an intelligent, self-adjusting, and highly optimized network. Companies that succeed in integrating these technologies will gain a significant competitive edge by creating resilience, reducing costs, and meeting rising expectations for speed, visibility, and environmental sustainability. Let us briefly look at these promising technologies and developments.

8.1.1 WAREHOUSE AUTOMATION AND ROBOTICS

Warehouse automation of the future shall undergo revolutionary changes driven by Artificial Intelligence, IoT, Edge Computing, Digital Twins, and Advanced Robotics. These technologies will enable decision-making in real-time, self-optimization, and bidirectional humans-to-machines interactions, in different contexts such as,

- Hyper-Automated Warehouses: AI-based control systems shall manage inventory, order fulfillment, and last-mile logistics with reduced intervention.
- Autonomous Mobile Robots & Swarm Robotics: They will collaborate efficiently, minimizing space utilization and travel paths in real time.

- Advanced Robotic Manipulators: AI-powered robotic arms will handle a wider variety of objects with dexterity, making automation feasible for diverse industries.
- Collaborative Robotics (Cobots): Robots will collaborate with humans, increasing productivity while minimizing operational risks.

The transformation of warehouses through automation and robotics is fundamentally driven by advanced engineering tools and mathematical techniques that optimize decision-making, coordination, and adaptability in logistics systems. Several critical domains contribute to this evolution, each grounded in mathematical principles that enhance efficiency and scalability such as Edge Computing & IoT, which enhance creation of digital twins, as key elements to support the AI powered decision support systems.

The convergence of edge computing and the Internet of Things enables real-time decision-making in distributed warehouse networks. In this paradigm, computational models leverage distributed optimization techniques, where decisions are made locally at edge devices rather than relying on a central cloud system. The information provided allows us to create a virtual representation of a physical warehouse, continuously updated with real-time data. The mathematical backbone of Digital Twins lies in dynamical system modeling and stochastic process simulation. The warehouse is modeled using discrete-event simulation, where the system state changes at discrete points in time, governed by probabilistic transition functions. Mathematically, a Digital Twin can be represented as a Markov Decision Process:

$$S_t = f(S_{t-1}, A_{t-1}, W_t) \tag{8.1}$$

where $S_t$represents the warehouse at time t, A_t represents control actions, and W_t represents stochastic environmental factors. Monte Carlo methods and agent-based modeling (ABM) further refine the system by simulating thousands of potential operational scenarios.

The future of warehouse automation lies squarely in mathematical optimization, AI-driven control, and decentralized coordination mechanisms. From Markov Decision Processes that dictate the flow of robots to Graph Theory that optimizes warehouse design, the intersection of these methods produces incredibly efficient, scalable, and robust logistics systems. As these engineering tools continue to advance, they will transform the global supply chain, allowing unprecedented levels of automation and efficiency. The challenge will be to develop this vision so that massive implementation can occur in the near future.

Real-time data aggregation enabled by IoT reduces stock errors, improving inventory accuracy from 80 to 90% (current automation) to 98 to 99% (IoT + Edge Computing). Using Bayesian updating for continuous inventory adjustment:

$$P(H|D) = \frac{P(D|H) \cdot P(H)}{P(D)} \tag{8.2}$$

where H represents a correct inventory prediction and D represents real-time sensor updates, the estimated value creation ranges,

- 5-15% reduction in stockouts and overstock scenarios.
- 3–7% lower carrying costs per year.

There are also benefits because traditional warehouses rely on scheduled maintenance with failure rates between 5 and 10% per year, causing unexpected downtimes. Predictive maintenance based on the Internet of Things reduces failures by 50–70%, reducing downtime and maintenance costs. Depending on the Failure Reduction Model (FR), reduction rates are feasible.

- Without IoT: $FR = \lambda \cdot e^{-\lambda \cdot t}$, with $\lambda \approx 0.05$ (5% yearly failure).
- With IoT: Real-time data adjusts λ dynamically, leading to failure rates of 0.015–0.025 (1.5–2.5%).

These reduction rates led to 15 to 30% of reduction in maintenance costs, as well as warehouse uptime increases from the current 92-95% to 96-98%.

8.1.2 BLOCKCHAIN FOR SUPPLY CHAIN TRANSPARENCY

Blockchain technology improves *traceability* in global trade, ensuring compliance with regulations and reducing fraud.

During the next five to ten years, blockchain technology will radically transform supply chain transparency by making possible a never-before level of security, trust, and efficiency in logistics and warehousing. At its core, it will be enabled by the ability to construct decentralized, tamper-proof, and completely auditable ledgers to trace every transaction, movement, and modification throughout a supply chain.

The evolution of blockchain in logistics will cease to be a matter of mere record-keeping but will include AI-driven decision systems, IoT networks, and cryptographic advances within a smart, autonomous digital platform for global trade (see Figure 8.1).

Mathematics will be the driving force behind this change, as it is the backbone of blockchain's immutability, consensus, and cryptography.

At its simplest, blockchain rests on a cryptographic hash function, a mathematical operation that hashes any input data to yield a specific, fixed-size string.

$$H(x) = SHA - 256(x) \tag{8.3}$$

where xx represents any transaction or data record, and SHA-256 ensures that even a minuscule change in xx results in a completely different hash output. This mechanism guarantees that once a transaction is recorded in a warehouse's blockchain, it becomes mathematically impossible to alter it without detection, securing the integrity of the supply chain.

Figure 8.1 Augmented reality in a last mile delivery process.

In the coming times, Zero-Knowledge Proofs (ZKP) will play an integral role in warehouse automation. This cryptographic protocol allows a subject to demonstrate a statement to someone else without disclosing the data on which the statement is based. In logistics, this would allow a supplier to demonstrate the accuracy of a shipment without revealing confidential prices or names of suppliers. The proof is mathematically based on probabilistic verification algorithms, enabling computational efficiency while maintaining privacy:

$$P(validtransaction|proof) \approx 1 \tag{8.4}$$

without revealing confidential details about the transaction itself.

To facilitate scalability, more recent consensus algorithms will replace traditional Proof-of-Work (PoW) mechanisms, which are computationally demanding and energy-greedy. Instead, Proof-of-Stake (PoS) and Byzantine Fault-Tolerant (BFT) consensus models will dominate, allowing warehouse nodes to confirm transactions in a decentralized network without excessive energy consumption. The mathematical concept underpinning BFT ensures that even when a part of the network is compromised, the consensus remains secure. It relies on the quorum condition:

$$3 \cdot f + 1 > n \tag{8.5}$$

where f represents the number of potentially malicious nodes, and n is the total number of participants in the network. As long as the majority of nodes remain honest, fraudulent warehouse transactions will be rejected.

With added supply chain security, multi-party computation (MPC) and homomorphic encryption will allow multiple stakeholders to process logistics data together without exposing proprietary information. Founded upon modular arithmetic and lattice-based cryptography, this strategy allows computation on encrypted data directly on warehouse data to accomplish analytics without privacy loss.

As threats from quantum computing arise, the supply chain sector will transition to resistant quantum blockchain protocols founded on mathematical functions that remain computationally infeasible for quantum attackers. Lattice-based cryptography, for instance, will provide a robust defense, utilizing the hardness of the shortest vector problem (SVP) in high-dimensional integer lattices:

$$\min||v||, \quad \text{where} \quad v \in L \quad \text{and} \quad v \neq 0 \tag{8.6}$$

ensuring that even future quantum computers will be unable to break the cryptographic security of warehouse transactions.

From a business perspective, blockchain's integration with automated smart contracts will eliminate inefficiencies in warehousing, enabling real-time execution of payments, shipments, and compliance verification without human intervention. These contracts, powered by deterministic finite automata, operate on a set of predefined states:

$$\delta(q, \sigma) = q' \tag{8.7}$$

where the contract transitions from state q to q' in response to an input σ, say, successful delivery of goods in an automated warehouse.

The impact of these innovations will be profound. In a supply chain network across the globe, every warehouse movement, inventory update, and order fulfillment will be recorded in an open, tamper-evident ledger, eliminating fraud, counterfeiting, and inefficiencies. Machine-based checks for compliance using AI-driven anomaly detection will flag suspicious transactions and provide regulatory and contractual compliance. Blockchain, AI, and high-security cryptography will build a supply chain logistics system where conflicts are nearly zero, audits are real-time, and trust is an inherent part of the supply chain structure.

As this new world unfolds, blockchain will no longer be a passive ledgering tool but an active participant in warehouse automation, optimizing operations in real-time and liberating trillions of dollars of economic value globally. The seamless convergence of decentralized math, autonomous AI, and quantum-resistant cryptography will define the warehouses of the future—smart, self-managing, and inherently secure.

Currently the optimization of last-mile delivery problem is an area for big improvement, and integrated traceability of the logistic process information is limited and fragmented. Users need to access to big carriers' information systems, when available, to collect some details of the process. This is an extra risk because of the service-driven exposition of IT systems.

8.1.3 ADVANCED MANUFACTURING AND SMART FACTORIES

With the era of smart factories and smart manufacturing, logistics will not be a second-order activity of just the movement of stuff from one place to another. Instead, it will be integral to the cyber-physical production system, changing flexibly with real-time conditions, choreographing material flows through smart automation, and seamlessly integrated with global supply chains. The very logistics definition in such a scenario is marked by hyper-connectivity, where indoor material handling, manufacturing activities, and external logistics systems together act as a unified system being powered by data, AI, and robotics.

In a smart factory, logistics is a real-time optimization problem, where the problem is not just moving goods from point A to point B but getting the right material to the right workstation, in the correct sequence, exactly when it is needed. It achieves this using a multi-layer network of autonomous mobile robots, automated guided vehicles, and intelligent conveyor systems that are all decentralized agents in an extremely coordinated network (see Figure 8.2).

The mathematical foundation of this optimization is in multi-agent reinforcement learning, wherein every robotic unit can learn and adjust to the dynamic factory

Figure 8.2 Collage showing different applications.

environment by altering its path planning, velocity, and task priorities. Movement of material through the factory will be achieved with shortest-path algorithm and dynamic flow optimization models such that the layout of the factory is represented by a weighted graph and graph-based reinforcement learning provides the optimal transportation paths.

Advanced manufacturing will integrate *cyber-physical systems* and *digital twins* to optimize production workflows. Smart factories leverage *real-time data analytics* and *predictive maintenance* to reduce downtime and increase operational efficiency. Companies such as *Siemens and GE* have implemented *Industrial IoT (IIoT) platforms* to enable adaptive logistics and smart inventory management.

8.2 CONCLUSION

The logistics sector stands at the threshold of revolutionary changes driven by technology innovation and evolving customer needs. Major sectors such as Artificial Intelligence and Machine Learning, Digital Twins, Blockchain, Quantum Computing, Autonomous Logistics, and sustainability will play a critical role in making logistics processes more efficient, transparent, robust, and sustainable.

Machine Learning and Artificial Intelligence will enable strong prediction capability, improved demand forecasting, real-time optimization, and more resilient supply chains through reinforcement learning and graph neural networks.

IoT and big data-driven Digital Twins will enable virtual real-time simulation of supply chains so as to be able to precisely model and optimize in advance of actual implementation. Integration with higher-level mathematical models like Petri nets and discrete-event simulations will assist operationalization of precision and responsiveness (see Figure 8.3).

Blockchain technology will yield unparalleled security and transparency of supply chains via use of cryptographic functions, zero-knowledge proofs, and consensus procedures, thereby ending fraud and ensuring compliance and tracing. Future

Figure 8.3 AI generated context for DT in the near future.

implementations of blockchain technology will involve use of quantum-resistant cryptography, increasing security against increasing computational danger.

Quantum computers will revolutionize logistic optimization through effective solving of current intractable problems within routing and scheduling using quantum annealing and quantum-inspired algorithms.

Autonomous logistics like autonomous trucks and robotic automation platforms will revolutionize warehouse activity, eliminating human error and operational costs and enabling greater responsiveness and flexibility with 5G and edge computing technologies.

Sustainability will remain at the forefront of thinking, with circular economy principles woven into logistics activities through AI-based multi-objective optimization models, considering economic efficiency, compliance, and environmental responsibility simultaneously.

In industrialized manufacturing settings, logistics is embedded in smart factory operations, enabling real-time coordination of decentralized robotic agents and intelligent systems through multi-agent reinforcement learning, shortest-path algorithms, and dynamic optimization models.

Together, the integration of these advanced technologies will lead to smart and highly optimized logistics networks with efficiency, security, sustainability, and resilience, which will produce substantial competitive advantages for organizations that embrace this technological shift [7].

360 Degree Review: Performance review that includes feedback from superiors, peers, subordinates, and clients.

ABM: Agent-based modeling.

Abnormal Variation: Changes in process performance that cannot be accounted for by typical day-to-day variation. Also referred to as non-random variation.

AI: Artificial Intelligence.

AMR: Autonomous Mobile Robots.

BTF: Byzantine Fault-Tolerant.

DES: Discrete-event simulation.

DT: Digital Twin.

GNN: Graph neural networks.

IoT: Internet of Things.

MDP: Markov Decision Process.

MPC: Multi-party computation.

PoW: Proof-of-Work.

SVP: Shortest vector problem.

ZKP: Zero-Knowledge Proofs.

Bibliography

1. Saurabh Agrawal, Rajesh K Singh, and Qasim Murtaza. Disposition decisions in reverse logistics: Graph theory and matrix approach. *Journal of Cleaner Production*, 137:93–104, November 2016. ISSN 0959-6526. doi: 10.1016/j.jclepro.2016.07.045. URL `http://dx.doi.org/10.1016/j.jclepro.2016.07.045`.
2. Wafaa AH Ahmed and Alexa Rios. Digitalization of the international shipping and maritime logistics industry: a case study of tradelens. In *The digital supply chain*, pages 309–323. Elsevier, 2022.
3. apax. Tosca to acquire polymer logistics — apax partners, 10 2019. URL `https://www.apax.com/news-views/tosca-to-acquire-polymer-logistics/`.
4. Donald J Bowersox. *Logistical excellence: it's not business as usual*. Elsevier, 2013.
5. Donald J Bowersox, David J Closs, M Bixby Cooper, and John C Bowersox. *Supply chain logistics management*. Mcgraw-hill, 2020.
6. Sustainable Business. Ups advances sustainable operations with big data—trellis, 2013. URL `https://trellis.net/article/ups-advances-sustainable-operations-big-data/`.
7. Martin Christopher. *Logistics and Supply Chain Management: Logistics & Supply Chain Management*. Pearson UK, fifth edition, 2016. ISBN 9781292083797.
8. Antonios Garas, Frank Schweitzer, and Shlomo Havlin. Ak-shell decomposition method for weighted networks. *New Journal of Physics*, 14(8):083030, August 2012. ISSN 1367-2630. doi: 10.1088/1367-2630/14/8/083030. URL `http://dx.doi.org/10.1088/1367-2630/14/8/083030`.
9. Gianpaolo Ghiani, Gilbert Laporte, and Roberto Musmanno. Introduction to logistics systems management. *Introduction to Logistics Systems Management*, 2 2013. doi: 10.1002/9781118492185. URL `https://onlinelibrary.wiley.com/doi/book/10.1002/9781118492185`.
10. Gianpaolo Ghiani, Gilbert Laporte, and Roberto Musmanno. *Introduction to Logistics Systems Management: With Microsoft Excel and Python Examples*. John Wiley & Sons, 2022.
11. David B Grant, Chee Yew Wong, and Alexander Trautrims. *Sustainable logistics and supply chain management: principles and practices for sustainable operations and management*. Kogan Page Publishers, 2017.
12. Roman Gumzej. Intelligent logistics systems in e-commerce and transportation. *Mathematical Biosciences and Engineering*, 20(2):2348–2363, 2022. ISSN 1551-0018. doi: 10.3934/mbe.2023110. URL `http://dx.doi.org/10.3934/mbe.2023110`.

13. Jessica Hanafi, Sami Kara, and Hartmut Kaebernick. Generating fuzzy coloured petri net forecasting model to predict the return of products. In *Proceedings of the 2007 IEEE International Symposium on Electronics and the Environment*, page 245–250. IEEE, May 2007. doi: 10.1109/isee.2007.369402. URL `http://dx.doi.org/10.1109/ISEE.2007.369402`.
14. Alan. Harrison, Heather. Skipworth, Remko I. van. Hoek, and James. Aitken. *Logistics management and strategy : competing through the supply chain*. Pearson Education Limited, 2019. ISBN 978-1-292-18370-1.
15. Jun He, Lingxiao Xu, and Bo Li. Optimization and evaluation methods for automation of e-commerce logistics networks. In *2024 16th International Conference on Computer and Automation Engineering (ICCAE)*, page 59–63. IEEE, March 2024. doi: 10.1109/iccae59995.2024.10569460. URL `http://dx.doi.org/10.1109/ICCAE59995.2024.10569460`.
16. Dmitry Ivanov and Alexandre Dolgui. Viability of intertwined supply networks: extending the supply chain resilience angles towards survivability. a position paper motivated by covid-19 outbreak. *International journal of production research*, 58(10):2904–2915, 2020.
17. Niloofar Jefroy, Mathew Azarian, and Hao Yu. Moving from industry 4.0 to industry 5.0: what are the implications for smart logistics? *Logistics*, 6(2):26, 2022.
18. Iqra Sadaf Khan, Muhammad Ovais Ahmad, and Jukka Majava. Industry 4.0 and sustainable development: A systematic mapping of triple bottom line, circular economy and sustainable business models perspectives. *Journal of Cleaner Production*, 297:126655, 2021.
19. John Mangan and Chandra Lalwani. *Global logistics and supply chain management*. John Wiley & Sons, 2016.
20. Jingyang Mao, Ying Sun, Xiaojian Yi, Hongjian Liu, and Derui Ding. Recursive filtering of networked nonlinear systems: A survey. *International Journal of Systems Science*, 52(6):1110–1128, 2021.
21. Wei Mu and HePing Ding. E-commerce intelligent logistics data based on neural network model. *Mobile Information Systems*, 2022:1–12, September 2022. ISSN 1574-017X. doi: 10.1155/2022/8993365. URL `http://dx.doi.org/10.1155/2022/8993365`.
22. Tadao Murata. Petri nets: Properties, analysis and applications. *Proceedings of the IEEE*, 77(4):541–580, 1989.
23. Michael Naor, Alex Coman, and Anat Wiznizer. Vertically integrated supply chain of batteries, electric vehicles, and charging infrastructure: A review of three milestone projects from theory of constraints perspective. *Sustainability*, 13(7):3632, 2021.
24. Chengyang Ou, Fubin Pan, and Shuangjiao Lin. Cascade failure-based identification and resilience of critical nodes in automotive supply chain networks. *Sustainability*, 16(13):5514, June 2024. ISSN 2071-1050. doi: 10.3390/su16135514. URL `http://dx.doi.org/10.3390/su16135514`.
25. H Chr Pfohl. *Logistics systems*. Springer, 2010.

26. Eric Reed. History of tesla: Timeline and facts. *The Street*, 2020.
27. David Simchi-Levi, Philip Kaminsky, and Edith Simchi-Levi. *Designing and managing the supply chain: Concepts, strategies, and cases.* McGraw-hill New York, 1999.
28. Matthew A Waller and Stanley E Fawcett. Data science, predictive analytics, and big data: a revolution that will transform supply chain design and management, 2013.
29. Yulong Xi, Fengming Tao, Jia Yang, and Schanelle Lison Brooks. Optimization of location-routing planning in logistics distribution network for e-commerce enterprises using dpc-dnspso algorithm. *SSRN Electronic Journal*, 2022. ISSN 1556-5068. doi: 10.2139/ssrn.4231907. URL `http://dx.doi.org/10.2139/ssrn.4231907`.
30. Xinyi Xia et al. Scm pma supply chain models—a case study of tesla motors. *International Journal of Frontiers in Engineering Technology*, 4(4):33–40, 2022.

Index

For Product Safety Concerns and Information please contact our
EU representative GPSR@taylorandfrancis.com Taylor & Francis
Verlag GmbH, Kaufingerstraße 24, 80331 München, Germany